R&D, INNOVATION AND COMPETITIVENESS IN THE EUROPEAN CHEMICAL INDUSTRY

R&D, Innovation and Competitiveness in the European Chemical Industry

Edited by

Fabrizio Cesaroni
Laboratory of Economics and Management,
St. Anna School of Advanced Studies, Pisa, Italy

Alfonso Gambardella
Laboratory of Economics and Management,
St. Anna School of Advanced Studies, Pisa, Italy

and

Walter Garcia-Fontes
Universitat Pompeu Fabra,
and Centre de Referència d'Economia Analîtica,
Barcelona, Spain

KLUWER ACADEMIC PUBLISHERS
BOSTON / DORDRECHT / LONDON

A C.I.P. Catalogue record for this book is available from the Library of Congress.

ISBN 1-4020-7866-8 (HB)
ISBN 1-4020-7942-7 (e-book)

Published by Kluwer Academic Publishers,
P.O. Box 17, 3300 AA Dordrecht, The Netherlands.

Sold and distributed in North, Central and South America
by Kluwer Academic Publishers,
101 Philip Drive, Norwell, MA 02061, U.S.A.

In all other countries, sold and distributed
by Kluwer Academic Publishers,
P.O. Box 322, 3300 AH Dordrecht, The Netherlands.

Printed on acid-free paper

Printed in the Netherlands.

Contents

Part 3:

POLICY REGULATION AND THE ENVIRONMENT

Preface

This book is the outcome of a two year project involving a group of European economists and business scholars. Why the chemical industry? There are many reasons. First this is an industry in which Europe has been competitive for a long time. It was natural to think of it as a test-bed for understanding the sources of its performance. Second, the chemical industry is ideal for scholars who want to understand the complexity of industrial phenomena. The industry is large and it encompasses many sub-sectors. The latter are heterogeneous, and this implies that the industry as a whole can be used as a case to understand several phenomena, ranging from R&D and innovation to market structure, government regulation, business strategy, economies of scale and scope at the level of firms or at the level of entire markets. In this respect, the book attempted specifically at "fixing" industry characteristics to look, within the context of the chemical sector, at a number of topic that help us better understand the economics of industries and firms.

The spirit of the book continues the tradition initiated with a previous long term research project on the sources of competitive advantages and long-term growth of the US chemical industry. We tried to reproduce, for the European chemical sector, the approach taken by Ashish Arora, Ralph Landau, and Nate Rosenberg in the earlier book. The US project/book hinged on the breadth and complexity of the industry to understand industry structure and dynamics from different perspectives. The strength and success of its approach provided a natural thrust for repeating it here. At the same time, beside the focus on Europe rather than the US, this book developed other topics that were not addressed by the earlier book, and presented some newer perspectives. But like the previous one, it highlights the virtues of collective research among groups of motivated scholars who provide their special skills, interests, and expertise in different areas to contribute to an integrated output.

This is also why our recognition of the support of the DG Science & Technology of the European Commission (TSER Contract N°SOE1-CT97) goes beyond a mere grateful acknowledgement. It was the nature and the spirit of the TSER grants that allowed groups of scholars from different European countries, and with different backgrounds, to come together, and accomplish a project that integrated their diversities of approaches, knowledge, skills and expertise. We can only wish that the Commission feels encouraged from outputs like ours that there is serious value in enhancing collaborative projects such as the one that we had the fortune to undertake. In this context, we would like to warmly thank Virginia Vitorino, who was our scientific officer on behalf of the Commission, for her clear guidance in all administrative matters but also for her scientific advise as well as for her permanent support and kindness. We would also like to thank Paul David and Ed Stenmueller, who were active participants in some of the seminars leading to the production of this book, and gave invaluable suggestions that improved the papers and the editorial process. We carried onto this book a number of the lessons that we received from Nate Rosenberg an Ralph Landau in the realization of the previous projects, and more generally as their students. None of the above can of course be held responsible for the contents of this book.

We finally acknowledge the patience of all our authors, and the enthusiasm with which they undertook this effort, going through several tedious revisions and demands from us. A lot of the credit for whatever good is in this book goes to them. We, as editors, will take all the blame.

Fabrizio Cesaroni
Alfonso Gambardella
Walter Garcia-Fontes

INTRODUCTION

Fabrizio Cesaroni[*], Alfonso Gambardella[*] and Walter Garcia-Fontes[**]
[*]*St.Anna School of Advanced Studies and* [**]*Universitat Pompeu Fabra-CREA*

1. OBJECTIVE

The chemical industry is one of the largest and most R&D-intensive manufacturing sectors in all advanced economies, and its innovative patterns and productivity growth processes can have profound impacts on economic growth as a whole (Arora, Landau and Rosenberg, 1998). The European chemical industry is part of this general framework. While European chemical firms have played a crucial role in the early stages of the industry – the modern chemical industry has developed in Great Britain in the second half of the Nineteenth century – and during most of the Twentieth century, some indications suggest that in the last decades Europe has started loosing grounds vis-à-vis the other advanced regions, especially in the high-tech segments. Hence, the main objective of this book is to develop a comprehensive analysis of the European chemical industry by studying the development and shaping of the industry, and its impact on other industries, different regions and small-medium enterprises (SMEs), while comparing its performance with the cases of Japan and the US.

The studies collected in this book represent the research results of the Targeted Socio-Economic Research Project "From science to products: A *Green*

F. Cesaroni et al. (eds.), R&D, Innovation and Competitiveness in the European Chemical Industry, 1-21.

Paper on innovation in the chemical industry" funded by the European Commission. The project aimed at exploring two main issues: (i) the science & technology base and the dynamic performance of the European "system of innovation" in the chemical industry; (ii) the forces that encourage the diffusion of chemical innovations on downstream user industries, across regions, and on SMEs.

Concerning the first objective, we aimed at understanding how the science and technology base in the chemical industry relates to innovation and market development, competitiveness, and economic growth in general. We devoted particular attention to the ability of the chemical industry in Europe to translate its research into commercially useful products, and the problems thereof. In pursuing this objective, we focused on issues like: the importance of historical factors in affecting the dynamics of this industry in the three advanced regions (Europe, the US, and Japan); the effects of differences in market structure on their performance; the nature of the "transmission mechanism" from upstream scientific and technological research on commercially useful new products and processes; the localisation of chemical research worldwide and the role of strategic alliances and mergers and acquisitions in shaping the present dynamics of the industry; the European competitiveness in the new R&D-intensive specialty chemical sectors; the effects of different financial systems on innovation performance.

These issues are clearly crucial in determining the innovativeness and competitiveness of single chemical firms and of the European industry on the whole. However, similarly important are issues related to the diffusion of chemical innovations. We devoted particular attention to the nature of the diffusion process (e.g. user involvement, user-producer interactions), and more generally to the factors that encourage greater spillovers from chemical research and innovation on other industries, regions, and SMEs (e.g. in fields like biotechnology, environmental technologies, specialty chemicals). In particular, we discussed issues such as: the effects of different forms of government interventions on the incentives of chemical firms to undertake research characterised by extensive spillovers; the effects of licenses and of the rise of a market for chemical technologies on the growth of other firms or industries; the potential benefits produced by the development of new technologies by the chemical industry (like environmental technologies) on other firms, and particularly on firms like the SME which cannot develop these technologies in-house.

Although most of the issues discussed in this book affect the chemical industry regardless of its geographical position, we specifically adopted a European perspective, and tried to assess how European chemical firms are performing compared to Japanese and US firms. However, concerning the geographical dimension of the industry, a special attention has to be paid. The chemical industry has always been a "global" industry, as shown by the rapid increase in the number of chemical plants built in foreign countries by European, American and Japanese companies (Arora, Garcia-Fontes and Gambardella, 1998). Proximity to customers in fast-growing regions, a higher product differentiation and customisation of plants, and the need to reduce transport costs are important factors explaining the internationalisation of this industry. In most sectors, what were national companies competing as oligopolists in predominantly national or regional (e.g., European) markets have become international companies competing against each other in the global market place. In each sector, competition is a matter of intense competition between six to twelve large firms that dominate the market on a global base. According to this view, analysing the "European" dimension of the industry might be misleading.

Nevertheless, the regional dimension seems still to be relevant. First, because besides large multinational companies competing on a global base, there are several smaller firms competing on a narrower regional or local base. The performance of the European chemical industry, expressed in terms of innovativeness and competitiveness, strictly depends on the flows of resources (e.g. knowledge, technologies, human resources) that move from the former to the latter and *vice versa*.

Second, because the comparative positions of the European, American and Japanese chemical industries reflect the evolution of these industries over time. There are similarities and differences in the evolution of industry structure and innovation processes in these three regions. Some of the historical factors that have shaped the evolution of the industry over time have preserved and accentuated initial differences in industry structure and innovation processes in the three regions, while others have induced greater similarity and convergence across the three areas. For instance, the rise of scientific capabilities and the increasing role of users have been of the first type, while international trade in goods and technologies have been of the second (Arora and Gambardella, 1998).

Therefore, we believe that, in order to fully appreciate the competitive position of the European chemical industry compared with that of Japan and the US, it worth recalling the main factors that caused the chemical industry to evolve. The evolution process has been driven by advances in technology and by those institutions that have facilitated the growth of new markets, such as markets for technology. In some sense, the evolution of the industry has been characterised both by the presence of a series of major *discontinuities at the industry-level*, and by a rather stable *continuity in companies' life*. On the one hand, the dyestuff model, the development of polymer chemistry (i.e., the science of chemical products), and the chemical engineering (i.e., the science of chemical processes) were major changes in the knowledge sphere. The shift from coal to petrochemicals in the years before the Second World War had strong consequences on regional leadership in chemicals, and allowed the American chemical industry to catch up with Europe. The emergence of specialised engineering firms (SEFs) made it easier the outsourcing of process technologies and allowed a growing division of labour at the industry level between SEFs and chemical companies. The world demand decrease during the 1980s induced a process of industry restructuring.

On the other hand, chemical firms were able to evolve and compete over time. BASF, Bayer, Dow Chemical, Agfa, Du Pont, i.e. some of the leading chemical companies nowadays, have more than one hundred years history and have been top chemical producers during all this period. This means that between small and large companies, markets, research institutions and other organisations there has been a process of co-evolution, with firms playing the central role within the chemical system.

Two factors that have arguably been the motive force for the growth of the chemical industry in the last half century lie on the technological domain: polymer chemistry, and chemical engineering (Landau, 1998). *Polymer science* has been the source of a large number of major product innovations, and is the basis for very many of the sectors regarded as the heart of the chemical industry, including synthetic fibers, plastics, resins, adhesives and paints and coatings. The other, *chemical engineering*, has been responsible for making possible the production of these polymer-based products (and polymer building blocks) at a cost low enough to ensure their success. The rate and direction of technical advance in these domains has changed in the last twenty years, and with it, the identities of the main actors have changed as well. These changes, which have taken place against a backdrop of slower economic growth in most of the developed world, have also contributed to significant changes in the strategies of many of the leading firms in the industry, and to the dramatic shifts in industry structure in recent years.

The rise of petrochemicals in the post-World War II period laid the basis for the technological convergence of the oil-refining and chemical sectors. Indeed, polymer chemistry showed how petroleum and gas-based feedstocks could be used to make very useful products. In addition, chemical engineering provided the basis for the design of both refineries and chemical plants for products such as polyethylene and nylon. The impact of this convergence in the oil and chemical sectors was asymmetric. While oil firms moved downstream into chemicals, chemical firms were much less successful in moving upstream into oil refining. Furthermore, oil firms retained control of crude-oil supplies, and had existing distribution channels and networks for by-products of oil-refining which were difficult for chemical producers to try to duplicate. As a consequence, chemical companies tried to react to their inability to move upstream by moving further downstream into differentiated products.

Following the big technology push in the industry during the 1950s and 60s, technology diffused more widely than it ever had before. Specialized engineering firms played a key role in creating a global market for process technologies for a large number of basic and intermediate chemicals. The maturing technology, along with increasing competition and slower demand growth, lowered the payoffs to traditional types of innovations. Commercialisation became more expensive and required ever more sophisticated knowledge of customers and the market. Faced with over-capacity, the industry restructured, beginning in the 1980s in the US, and a few years later in Western Europe. The drive to reduce cost dominated the initial restructuring phase, driven in part by the relentless pressure from shareholders and their representatives. Major realignments of the product portfolios of many firms followed, with many mergers and acquisitions and the rise of entirely new firms in the industry.

During this phase, many firms cut down on R&D and refocused R&D expenditures on short term projects and away from more fundamental research. In the past couple of years, there are some indications that the industry may be entering a new phase of technological change and R&D spending appears to be picking up as well. Nonetheless, the restructured firm portfolios beg the question of who will perform the basic research that continues to be very important for the future of the industry. The current situation points to the possible need for increased government support for R&D in an industry that has hitherto largely financed its research by itself. This book aims at offering a tentative answer to these questions.

2. HOW THIS BOOK IS ORGANISED

Part 1: Market structure and innovation

The first part of the book explores the relationship between market structure and the rate and direction of the innovation process. Specifically, the chapter by Marin and Soitis analyses the relationship between market size, the existence of alternative technological trajectories in R&D intensive industries, and concentration. In particular, according to Sutton (1991 and 1998) predictions, Marin and Siotis tested two main hypotheses: a) as market size grows, industries with endogenous sunk costs, such as advertising and R&D expenditures, may not evolve in equilibrium towards fragmented market structures; b) high R&D intensity industries' minimum concentration levels are affected by consumers' heterogeneity of preferences, since the latter fosters the existence of alternative technological trajectories. In that context, the heterogeneity in research trajectories yields more fragmented structures.

These two hypotheses can be correctly tested in the case of the chemical industry, which is formed by many sectors where different firms are operating. Each sector is characterised by a different R&D intensity and a different number of product classes, and presents a different market structure. In none of these sectors advertising expenditures are particularly relevant since most of their products are sold as intermediate outputs.

By using a specifically defined database, the two hypotheses have been tested. One of the key result suggests that greater fragmentation of the industry implies that research efforts devoted to single product lines generate fewer spillovers for products in other lines. In turn, in R&D intensive industries, the degree of fragmentation reduces the degree of concentration. Indeed, when investing in one research trajectory, one firm may reach a high market share only when some conditions are met. This occurs when there is a high substitutability among varieties of the product and all consumers start buying the superior quality variety. Alternatively, there are strong scope economies and innovations can be applied to all product varieties made by the firm. In R&D intensive industries substitutability among varieties plays its role. Firms in the industry produce several groups of products that are imperfect substitutes in consumption and on the supply side, since they embody different technologies. In this case, R&D expenditures devoted to one group of product have limited spillovers on another group of products. Firms can develop many different technologies, each one related to a different group of products and must choose either to spend all their money on one trajectory or to distribute it among several trajectories.

In the subsequent chapter, Ornaghi, Marin and Siotis examine some general patterns of mergers and acquisitions (M&A) in the chemical industry between 1985 and 1997. They focus their attention on the investment strategies of US and European firms, both domestic and international. They concentrate on similarities and dissimilarities across the Atlantic with respect to domestic (intra-regional) and international M&As, and provide stylised facts on investment flows in order to identify differences across target areas and industries. This exercise is carried out by focusing on the type of industries involved, percentage of ownership acquired, and modes of entry. Besides, they provide an analysis of the mode of entry and the extent of ownership control.

As emphasised by Arora, Gambardella and Garcia-Fontes (1998), there has been an increasing globalisation process for this industry and M&A are the main tool used by parent companies to expand. The picture that emerges is that Europe and the US are the main source and destination of investments through mergers and acquisitions, while Japanese companies play appear to play a more modest role. Another interesting pattern that emerges is that, apart from high-tech sectors, there are no remarkable differences between domestic investments and foreign direct investments undertaken by industrialised countries in terms of the industries involved. In fact, the sectoral distribution and number of M&As undertaken by US and European firms is almost perfectly symmetric. This suggests that oligopolistic rivalry is an important driving force behind M&As in the chemical industry. Furthermore, as far as the mode of entry is concerned, a marked pattern is the proliferation of vertical acquisitions.

The last chapter of this section of the book focuses on corporate restructuring processes. Arora, Ceccagnoli and Darin provide a novel approach to the existing literature on the effects of restructuring on R&D investment by focusing on a single industry, chemicals. Indeed, the chemical industry is very research intensive and has experienced thorough restructuring since the early 1980s. By focusing on a single industry the authors were able to identify the technological and R&D features of its segments. This issue is particularly relevant, since there is evidence that restructuring affects R&D differently in businesses with different technological features. However, no study so far has provided a systematic inquiry into this link. Using a panel of 535 European, American, and Japanese firms for the years 1987-1997 they find restructuring to be an important component in the observed changes in R&D intensity. The study shows that restructuring affects R&D both through changes in size and through changes in the composition of business portfolios, and that these effects differ across industry segments.

Part 2: Technology development and diffusion

The second part of the book explores the nature of R&D processes, the development of new technologies, and the ways in which technological knowledge and assets are transferred and diffused. These topics have been analysed either for the chemical industry on the whole, or for specific segments, such as bio-pharmaceuticals or paints and inks.

In his study, Brusoni aimed at comparing the EU and US coating industries with respect to a specific case of environmental innovation: the adoption of waterborne formulations. Two case studies were designed that focused on the two largest segments of the coatings industry: decorative paints and vehicles paints. Particular attention was paid to the relationship between technical change and the opportunities and constraints posed by environmental regulation. The results of the analyses show that, from a technological point of view, the industry seem to be perfectly capable of coping with the diffusion of new, solvent free formulations. The diffusion of waterborne is then unlikely to generate an industry shake out. Incumbents seem to be well equipped with the R&D, manufacturing and marketing capabilities to pre-empt any challenge from potential entrants. Indeed, the capability to manage the interdependencies between R&D, manufacturing and marketing is the key factor that makes the paint market hardly contestable by new entrants.

Rather than a lack of technological capabilities, customer demand is the key to understand some of the problems related to the diffusion of waterborne products, particularly in the decorative business. In both the cases analysed, the role of demand is paramount. Whether 'naturally' different, or regulation induced, customers have played a fundamental role in driving (or constraining) the diffusion process of alternative formulations. Given the relative stability of the industry knowledge base, it is hard to consider technical capabilities as the key issue that set apart EU and US producers. The relevant factors have probably to be sought in the external environment, in the characteristics of demand and customer preferences, in the different role played by process and product regulation. Further, different financial settings may have facilitated the rationalisation efforts of US companies, in the sense that US financial investors have given a premium to the rationalisation and refocusing efforts greater than that granted by European investors.

The chapter by Mariani explores the geographical dimension of the R&D processes in the chemical industry. Specifically, her work traces the geographical distribution of R&D activities in Europe in 5 chemical sectors – i.e. biotechnology, pharmaceuticals, organic chemistry, materials and polymers – and explores research collaborations at the level of the individual inventors. It compares the firm and the geographical cluster as two organisational modes for producing interdisciplinary patents and for pulling together large teams of inventors. The results confirm that chemical companies perform most of their research in the home-country, and that patenting activity clusters geographically. However, compared to the geographical cluster, the large firm is a more effective mechanism for producing interdisciplinary innovations and for fostering the formation of large and de-localised teams of inventors.

The issue of R&D activity is further explored in the chapter by Geuna, Llerena, Matt and Savona, which explores the nature of university-industry collaborations in the chemical industry. Their study examines the evolution of contractual relationships in chemistry at the University Louis Pasteur of Strasbourg during the 1990s. Using entry, exit and persistency indicators they analyse the changes in the contracting behaviour of the various university laboratories and compare the types of contracts signed by persistent and non-persistent laboratories, persistent laboratories being those that had contractual relationships throughout the period. Four main conclusions emerge from this study. First, an increasing number of new actors (firms and university laboratories) have become contractual partners. Second, persistent laboratories are the most active actors inside the university. Third, the increasing number of contractual agreements signed by persistent laboratories underlines the existence of an ongoing management learning process. Finally, there is no clear-cut difference in the type of contracts signed by persistent and non-persistent laboratories indicating that the development of university-industry relationships is demand-driven. These results are framed within the debate on the ongoing changes in the role and behaviour of universities.

Finally, the chapter by Arora, Gambardella, Pammolli and Riccaboni focuses on the dynamics of the network of collaborative agreements in R&D in the bio-pharmaceutical industry after the "molecular biology revolution". The chapter is based on an empirical analysis that has been performed by using a comprehensive and specifically designed database. The analysis focused on the network of collaborative agreements in R&D and on license contracts relating to molecules under development drawn up among the firms of the sample still alive at the end of the period.

The results of this study show that, while the size of the network increases over time due to net flows of entry, its topological properties remain relatively unchanged. The evolution of the network has occurred without relevant deformations in the core-periphery profile. As far as age-dependent propensity to collaborate is concerned, the study found that the extent of inter-generation collaboration is much more significant than intra-generation collaboration. In addition, the propensity of firms of a given generation to enter into collaboration with firms of a different generation increases with the distance between the two, while the total number of intra-generation collaborations decreases over time and, moreover, tends to decrease for most recent generations. The study presented a unitary and coherent explanation of the evidence, coming to reveal the existence of a striking homology between structural properties of the dynamics of knowledge and of the evolution of network structure.

Part 3: Policy regulation and the environment

The last part of the book explores the environmental issue, the related political concerns, and the ways in which chemical companies modified their behaviour in order to respond to environmental questions. The first chapter of this part is by Arduini and Cesaroni. In their study, they explored the contribution of the European chemical industry to the development and diffusion of environmental technologies. The analysis has been primarily performed by using EPO and USPTO patent data and case studies.

Results of this study show a fairly higher innovative performance for Germany and the US, which seems to be linked to the different government regulation and public pressure that they face. As a matter of fact, the United States have faced environmental problems through very strict standards, and Germany has adopted the most rigid standards of Europe. Moreover in these countries, the public opinion have played an important role in influencing the environmental policy and behaviour of firms. These results could bring about the consideration that rigid environmental standards and strong public pressure have a positive influence on the environmental innovative rate. The role of chemical companies is key in this respect. Both US and European chemical companies patent more in clean technologies rather than end-of-pipe ones, hence confirming the efforts made by the chemical industry in reducing pollution of chemical processes.

At the same time, their case study analysis shows that firms' innovative behaviour has to satisfy economic constraints, and that environmental issues are conditioned by the search for technology efficiency and effectiveness. When the government regulation defines precise standards, and imposes precise prohibitions, these are clear targets that have to be considered in the innovative activity. Hence, innovative problems encountered by firms are not within the regulation itself, but they are related to the time needed to satisfy the standards according to the regulation.

Similarly, the chapter of Becker and Engelmann explores the environmental issue, but from a narrower perspective. The objective of their chapter was to investigate whether large firm size is conductive to a higher level of inventive activities in the field of water benign process technologies in the German chemical industry during the 1990s. By using patent applications as indicator for inventive activities, their study indicate that with respect to both end-of-pipe and recycling technologies especially very large firms including BASF, Bayer and Aventis (former Hoechst) predominate invention. Except Solvay and Henkel within the portfolio of patent applications of very large firms, patent applications in the field of end-of-pipe technologies play a less important role than patent applications in the field of recycling technologies. Hence, the most important patent applicants focus on a more sustainable technological pathway in protecting the environment during the 1990s.

The study clearly shows that, in taking their innovative decisions, firms take into account both internal and external factors, namely the existing or expected environmental regulation, and the need to maintain a high productivity level. Hence, firms favour to invent and implement recycling technologies at the expense of end-of-pipe technologies as governmental emission standards get tighter and tighter. Indeed, the ongoing addition of end-of-pipe technologies due to tightened standards during the 1990s in Germany requires ongoing material input and energy consumption while the incumbent production process remains unchanged. Thus, the implementation of (newly invented) end-of-pipe technologies tend to reduce productivity and thereby competitiveness.

3. POLICY CONCERNS

The considerations reported in the different studies collected in this book can be discussed in terms of policy implications, by specifically looking at the innovation- and research-based competitiveness of the industry. In order to do so, however, it is worth keeping in mind three preliminary remarks. First, the chemical industry is a quite complex and differentiated realm, and it is hard to focus on specific aspects, as the industry is composed of many different agents, market structures, and technologies, which makes it hard to discuss a unique set of policies or initiatives.

Second, the competitive position of the European industry is a difficult concept to incorporate into any policy analysis because of the international nature of the industry. US firms conduct large amounts of research and development in Europe and European firms conduct large amounts of research and development in the US. Nevertheless, different studies reported in this book have looked at the advantages or disadvantages of European firms vis à vis their US or Japanese counterparts.

Third, although we aimed to focussing to the European dimension, the level of internationalisation of the industry makes it difficult to think of local policies. Its increasing globalisation suggests that even national policies may be confined to a fairly narrow territorial level. The European dimension is in many respects the right dimension for innovation policies in this sector.

Hence, the discussion on the policy implications can start by analysing the sources of competitive advantages for the European chemical industry, with particular attention to the innovation-related sources. These sources can be found on the characteristics of the public research, on the role played by the larger firms, in the integration between large firms and public R&D, and in the role played by the smaller firms. Each of these sources can be analysed in terms of policy concerns.

3.1 Public research

One the possible sources for competitive advantage is based on the knowledge base from which firms in Europe may draw for innovation. We looked at this issue from different perspectives. On the one hand, some of the studies reported in this book have looked at the publicly funded research and tried to assess the performance of European public research institutions. This was done by looking at publications and patents and by trying to develop performance indicators. On the other hand, the importance of the knowledge base for specific subsectors was analysed.

One way to look at the contribution of public research is to assess the importance played by publications and technical reports in the process of knowledge transfer. For example, in a recent work, Geuna (2000) has examined the evolution of scientific specialization in the chemical and pharmaceutical fields in the four largest European countries (the UK, Germany, France and Italy), the EU as a whole, the US, and Japan. His results indicate that of the countries analysed the US has a much clearer specialization profile, both in terms of positive and negative specialization and in the area of research, indicating a much higher degree of knowledge integration than in the EU. The US has also the most stable publication profile; compared to the other countries and has the highest level of knowledge persistence. In general, when a country has a specialization in a scientific field in the area of basic research one can expect high persistency in that specialization. Furthermore, public research carried out in North America is valued and used extensively by the largest R&D firms in the pharmaceutical sector in the EU. This is consistent with the fact that the US has a persistent specialization in Medical Chemistry and Pharmacy & Pharmacology both in applied and in basic research.

Hence, despite the fact that in terms of number of publications and patents the European chemical innovation sector is performing well, according to other indicators, such as specialization and persistence, the US may be outperforming the EU. In various case studies, these general conclusions have been tested against the knowledge base of different industries.

In the case of the pharmaceutical industry, European countries exhibit capabilities in terms of applied and engineering research, but not in basic research. Such lack of basic research capabilities may well explain the frequency with which European R&D managers in pharmaceuticals approach the US knowledge base (Brusoni and Geuna, 2003). As for chemicals, the pattern of sourcing is different. As their home country knowledge bases seem more capable of providing a more integrated pattern of research capabilities, European chemical firms rely chiefly on their home country knowledge base and then approach the European one. They hint at the possibility that governments can actually influence the rate of technical change by fostering the development of an 'integrated' specialisation profile.

Similarly, the chapter of Brusoni within this book analyses the case of a more stable industry, the paints, coatings and printing inks industries. Contrary to scientifically more dynamic industries, in this industry European firms seem in a better position. For instance, from a technological point of view, the industry seems to be perfectly capable of coping with the diffusion of new, solvent free formulations. Brusoni does not spot any difference between US and European firms, as they all market waterborne products alongside traditional solvent based paints.

In sum, concerning the capability of the European chemical industry to translate its knowledge base into commercially successful products, it seems that, with respect to some indicators such as specialization and persistence, the knowledge base of the European innovation system does not perform as well as previously thought, so in some sense there may be fewer discoveries to push commercial applications than what is needed. Furthermore, from the evidence gathered from the pharmaceutical industry, it seems that some European firms are successful in translating scientific discoveries into products, but these discoveries are not based in Europe but mainly in the US.

3.2 Innovation policies and the large chemical firms

Large integrated firms play an important role in the European chemical industry. It is therefore important to understand the extent to which such firms need specific policy interventions to enhance their competitiveness and their ability to promote innovation, employment, etc.

Some of the key findings of the research project reported in this book about large firms can be summarised as follows.

Market structure. The chapter by Marin and Siotis analysed the market structure of the chemical industry using the framework developed by John Sutton (1991 and 1998). The goal of this approach is to predict the concentration of an industry from its R&D intensity and the degree of homogeneity of its products. The theory says that in high R&D intensive industries, when the product differentiation of the goods is low, one will observe concentration. The rationale is that a single firm can internalise a fair amount of spillovers since R&D targeting a given product can produce results that can be employed for developing the other products as well. R&D opportunities in one area can then be used to move into related areas, thereby increasing the concentration of the industry. By contrast, when product differentiation is high, and the various sub-markets are fairly independent from one another, opportunities in one product segment will not translate into analogue opportunities in other segments which are close in product space. As a result, a given industry which encompasses fairly differentiated products, will be concentrated if it was concentrated to start with; it will not be concentrated if it was originally composed of independent firms in each of the independent sub-markets.

Marin and Siotis' study used data on several segments of the chemical industry. The goal was to estimate whether R&D intensive segments implied higher concentration when there was a high degree of homogeneity of the products within that segment. They cast their analysis in terms of a test of Sutton's theory. From a policy perspective, their analysis can also be interpreted as an attempt to understand whether we should expect major changes in the structure of the main segments of the European chemical industry in the near future. Suppose that they found that high R&D intensive segments with high degree of product homogeneity were not concentrated. By using Sutton's framework, one would expect that these segments may undergo notable changes in industry structure. This is because some firms in such low concentrated segments may make a discovery in their product domain that may in turn enable them to "escalate", as Sutton puts it, into related segments. Marin and Siotis use extensive data on chemical industry segments in Europe, which makes their analysis particularly apt and complete to address these issues.

They find that the structure of the European chemical industry conforms quite well with the predictions of Sutton's theory. Particularly, in R&D-intensive segments the industry is concentrated. This suggests that one is unlikely to observe major shake-outs or relevant changes in most segments of the European chemical industry in the near future. As far as policy is concerned, this suggests that no particular action should be undertaken, or one is expected to undertake in this domain. Also, major crises in the industry (e.g. business failures, significant competitive threats, etc.) are unlikely to occur, with implied no anticipation of strong policies in this and related areas of intervention.

The organisation of R&D in the large chemical firms. The chapter by Mariani focuses on the organisation of R&D in chemical companies. Particularly, the study compares the type of R&D and the research networks promoted by the large chemical firms in Europe, and compares them with the type of R&D and research networks observed in geographical clusters. In so doing, the analysis attempts to compare large firms and geographical clusters as alternative modes for conducting R&D.

The first important result of this analysis is that the European chemical companies perform most of their research in their home-country, and that patenting activity clusters in few regions. This confirms earlier study by Patell and Pavitt (1997) that the globalisation of R&D by multinational enterprises is at best a quite incomplete process. The second set of results explore research collaborations at the level of the individual inventors. Particularly, the analysis compares the firm and the geographical cluster as organisational modes for giving rise to larger networks of inventors and for producing more interdisciplinary patents. The results indicate that, compared to the geographical cluster, the multinational company is a better mechanism for creating larger networks, for enhancing collaborations amongst de-localised inventors, and for producing interdisciplinary patents. In short, this confirms that the firm, and particularly the large companies, typically promote larger research networks, and they produce rather general sort of research, at least in the chemical business.

But this also suggests that, as far as the large European chemical firms are concerned, there is no urgent policy intervention for promoting the generation of R&D and related activities. The large European chemical firms do engage in these activities, and as a matter of fact they do give rise to large networks of inventors and they do produce patents with wide potential applicability.

International Investments. Another critical area for the competitiveness of the European chemical industry is the degree of internationalisation of its firms. It is well known that globalisation is becoming a key strategy for the competitiveness of companies in many industries, and this is especially true of the chemical industry. The main players in this industry, and particularly the larger firms, typically operate on a global scale. As a result, the extent to which the European chemical firms are globalised, particularly in comparison with their US and Japanese competitors, is key for understanding their competitive position.

As it has been demonstrated by recent studies (Arora, Gambardella and Garcia-Fontes, 2000) the chemical industry has become more global, with a lower share of plants of companies from one region located in the same region, and a higher share of plants from one region in the other regions. Furthermore, the European companies have proved to be particularly active in this globalisation process, as they have typically increased their share of plants abroad. Moreover, this happens both in advanced markets like the US and Japan, and in the open market of the developing countries, and particularly in Asia.

These results then suggest that the competitiveness of the major European chemical has not declined in the past decade or so. In turn, this confirms our previous remarks that no major policies are needed today for enhancing the ability of these companies to internationalise. It appears that they are continuing a long standing tradition of internationalisation, and there is no need for investing major policy resources in promoting patterns that are mastered quite effectively by these companies without any particular policy support.

R&D, innovation and restructuring in the chemical industry. The chapter by Arora, Ceccagnoli and Da Rin looked at how restructuring affects R&D, and they find that indeed restructuring does matter for R&D investment. Net acquisitions in R&D intensive industry segments have a positive and significant effect on R&D investment, a result robust to different specifications and samples. Amongst other things, they find that financial variables like debt or cash flow do affect R&D, but not the effect that restructuring has on R&D. They also separate the impact on R&D through changing size distributions due to restructuring from the direct impact. In other words, they study how much of the change in the average R&D intensity within industry segments is due to changes in scale distribution. The study finds restructuring to be an important component of the observed changes in R&D intensity. Moreover, the impact of restructuring differs across segments. For instance, in Life Sciences, most of the impact is through restructuring of firm portfolios rather than changes in the size distribution. In Other Chemicals, most of the impact is through changes in size distribution, with the size distribution becoming more equal after restructuring has had place. In Commodities, both matter, with an increase in size inequality as well as a direct increase in the inequality in R&D due to restructuring of the firm portfolios. These results provide a new, more composite, perspective on the effect of corporate restructuring on R&D.

The major policy implications from this analysis is that policy should encourage restructuring processes in Europe, especially because – as this study finds – they ultimate produce an increase in the R&D intensity of firms. Related to this is the fact that restructuring often involves serious costs as it implies changes in the structure of firms which entail short-run costs, both private and social (e.g. layoffs, reduced profits). These have often restrained restructuring processes. We suggest that important policy interventions are necessary in order to: a) reduce the social costs involved; b) separate them from the restructuring and other private costs of the companies.

In short, we suggest that governments should take up the short run social costs of restructuring that are often borne by individuals who are less capable of smoothing out incomes and losses over time (particularly in the short-run), or who would not be the same people that will benefit from the positive outcomes of company restructuring in the longer period. The private cost of restructuring should instead be borne by the shareholders, or whoever will benefit from the restructuring in the longer run. This separation is central to enable the restructuring especially in Europe were vested interests and political economy considerations are often involved in processes like this. In this respect, by separating social and private costs, and relatedly by attributing the costs to those who enjoy the longer run benefits of restructuring, we believe that restructuring processes will be easier to accomplish.

At the same time, governments should avoid intervening in the restructuring process, and on how it is carried out. The European experience of restructuring in the chemical as well as in other industries during the 1980s is that governments managed the restructuring process to a good extent, especially in France and Italy. The restructuring process should be governed by market forces. Moreover, the results of this study suggest that business swaps (e.g. exchange of divisions) by established firms through mergers and acquisitions, divestures, and the like, which have been often observed in recent years amongst the largest European chemical firms, should be encouraged.

On many occasions, these transactions have re-organised the underlying specialisations amongst leading firms by consolidating similar businesses that were in different firms. This improved the overall specialisation of the firms. The improved performance might provide one explanation why firms could employ new resources for additional investments in R&D, as the study by Arora, Ceccagnoli and Da Rin finds. Clearly, policy should also monitor these processes attentively to avoid that they create undesired monopolisations of certain industry segments. In short, policy should constantly check the balance between restructuring that consolidates and improves the specialisation of firms, and the coordination of certain businesses within one company, and the potential of these business swaps for anti-trust interventions.

Integration between large firms and public R&D (university)

For some industries, such as pharmaceuticals, large European firms have a less transparent access to public research than their US counterparts. This may be related to the fact that the intensity of R&D in some European industries is larger than the equivalent US industries. Another important factor that has to be taken into account is that firms in the chemical industry increasingly draw from a globalised knowledge base, benefiting from public R&D done internationally. In some industries there is a lack of technological diversity in the upstream knowledge base between large firms producing similar products. This would indicate that differences in competitiveness are not due to differences in the ability to acquire the expertise to patent in a given set of technological fields.

Summing up, the main conclusions from our analysis of public research and large firms indicate that large European chemical have little difficulty in obtaining technology. The differences in performance between European and US firms, or between different firms in Europe, should be sought not in differences in the ability of firms to obtain technology but in differences within firms in how they manage very similar technology. As a consequence, the key to understanding the competitive position of the European industry is understanding the differences between firms in how they manage technology, as it is this that determines relative economic performance.

Innovation policies and the small firms

Apart from leading world wide chemical firms, smaller companies play an important role in the chemical industry as well. Compared to large firms, smaller firms need greater policy support, at least in terms of the creation of suitable conditions for their growth, and for enhancing the potential that their growth may have for the evolution of the industry (especially the new high-tech segments) and its effects on competitiveness and employment.

First and foremost, the role of the smaller companies is linked to the opportunities for the development of a full fledged markets for technology in Europe, and more generally for the participation of the European firms in the global market for chemical technologies.

The chemical industry pioneered the growth of a market for chemical processing technologies. Since World War II, SEFs sold technologies through licenses to established chemical producers. This occurred from the US to other advanced countries (Europe and Japan) first, and in more recent years SEF from the US and Europe sold technologies to the developing countries. The main features of these markets are:

i) markets for technology allow for a significant diffusion of technology, which increases the investments of the companies operating in the final markets. Amongst other things, they reduce barriers to entry, as they make technologies available to companies with lower in-house technological assets, which in turn implies increased competition;

ii) these markets enable the formation of companies that specialise in the development of the technology even when they do not have the proper assets to develop and commercialise the final products. These companies are now formed because they realise that they can enjoy rents from selling the technology rather than having to gain such rents only through the sale of the final products in the much harder and competitive final markets;

iii) the formation of these companies has numerous advantages. First, they enhance the aforementioned process of technology diffusion, and they are the main vehicle for reducing technological barriers to entry, with implied increase in competition. Second, they induce other established producers to license because as technologies diffuse there is no advantage in trying to keep the technology secret; rather, the established producers themselves try to earn some rents in the market for technology, thereby enhancing the process of technology diffusion.

Apart from the SEFs, similar technology specialists have developed today in the biotech industry (see the chapter by Arora, Gambardella, Pammolli and Riccaboni), and in the environmental technology industry (see the chapter by Arduini and Cesaroni). Not only does this suggest that the phenomenon is more diffused than one could have originally thought, but also that many industries, and particularly many segments of the chemical industry are increasingly being organised in this fashion. Moreover, our analysis found that these technology specialists are less common in Europe than in the US, and that in Europe they deal with less advanced technologies (e.g. end-of-pipe rather than clean technologies in environment). In short, the market for technology appears to be less developed in Europe than in the US. While our analysis focused on the chemical industry, there are reasons to believe that the same applies to other industries as well. In turn, this suggests that the potential for technology diffusion and related economic benefits, are less pronounced in Europe.

This calls for adequate policy interventions that would remove the barriers to the creation of these markets. In some sense, one might say that the large chemical companies have a long standing tradition and competitiveness, which we confirmed with the various studies of this project. This implies that no major policy action is needed for enhancing their competitiveness. By contrast, European markets for technology are far from being developed, and this requires policy support for their formation. In particular, this calls for policy actions to encourage the rise and growth of smaller firms specialised in the development of technologies.

Specifically, we envisage the following policy actions for enhancing the markets for technology in the European chemical industry, and particularly in its engineering and technological sub-sectors:

- *Development of proper forms of Intellectual Property Rights (IPRs) to support the activities of smaller technology-based companies.* Our research found that IPRs can be an important factor for enabling such smaller companies to be founded and grow. This is because unlike the larger firms, they have no other means (e.g. downstream assets) for appropriating their innovations. As a result, IPRs can be the only form that they can employ to enjoy rents from their investments in research. The development of adequate, European wide forms of IPRs can then be critical for the growth of markets for technology in Europe.
- *Development of adequate forms of financing for new technology-based companies.* Apart from IPRs, markets for technology require new forms of financing for the small, technology-based firms to arise and grow. These include in particular forms that take into account the fact that these firms face a substantial technological risk compared to other activities. Hence, new forms of financing like venture capital should be strongly encouraged. While these issues have been emphasised on many occasions, here we provide a new perspective

about why such forms of financing can be important – notably, they encourage the growth of markets for technology, which implies reduced barriers to entry, greater technological diffusion, and new patterns and opportunities for economic growth.

- *Development of new forms of technology diffusion by universities, and the scientific institutions more generally.* Licensing by established producers increases when there are other agents that license. Moreover, institutions with no stake in the downstream markets have the highest incentives to license because they have nothing to loose in the downstream markets if new competitors arise. Universities or other research centres are a quintessential example of such institutions. As a result, by encouraging the diffusion of technology by universities (either directly or through spin-offs) there is an additional effect beyond the very diffusion of these technologies – notably that established producers in that technological domain will also be encouraged to license.

Finally, we want to emphasise that these policy implication are also important for the vertical structure of the chemical industry. Most notably, the rise of markets for technology implies a division of labour which in turn benefits the downstream producers. The classical advantages of a division of labour are indeed that the downstream producers can take advantage of the input at lower costs than if such input had to be produced in-house. Apart from efficiency gains in the downstream industries, this implies greater diffusion of the technology downstream, greater entry of new competitors in final markets, etc.. For example, SEFs imply efficiency gains, greater investments, and greater entry by chemical producers in downstream markets, and especially by producers that would not have been able to enter if they had to develop their technology in-house. Similarly, specialised engineering firms can be quite important in diffusing new environmental technologies. These advantage of vertical specialisation suggests some further policy actions, and particularly:

- Policies that would encourage the external monitoring of new technologies by existing producers in final markets, and more generally policies that would reduce the transaction costs for technology exchange that may exist in such markets. Transaction cost reducing mechanisms may range from the establishment of proper standards for reducing the potential segmentation of technologies that are in fact used for similar purposes, to the creation of proper institutions for technology exchange (e.g. standard contract).
- Policies that would reduce the search costs for new technologies, by creating new forums, electronics exchange markets, and the like for the exchange of technologies.
- Policies that would discourage the so-called "not invented here" (NIH) syndrome, which affects many firms and even countries. The NIH syndrome is

the one in which firms disregard technologies not developed internally, and it has been widely documented in the managerial literature, as we also noted in our research. As markets for technology develops, the problem with such a syndrome is that firms may loose important opportunities for acquiring technologies at lower costs, with implied benefits on their demand for technology, and ultimately on their profitability, competitiveness, and on their ability to increase employment. At the same time, such a syndrome may even prevent the markets for technology to arise in the first place, because the general business climate does not encourage the exchange of technologies amongst different parties.

4. REFERENCES

Arora A. and Gambardella A., 1998, "Evolution of Industry Structure in the Chemical Industry", in Arora A., Landau R. and Rosenberg N. (eds.), Chemicals and Long Term Economic Growth, John Wiley and Sons, New York.

Arora A., Gambardella A. and Garcia-Fontes W., 1998, "Investment Flows of Large Chemical Companies", University "Pompeu Fabra", Barcelona (Spain), mimeo.

Arora A., Landau R. and Rosenberg N. (eds.), 1998, Chemicals and Long Term Economic Growth, John Wiley and Sons, New York.

Brusoni S. and Geuna A., 2003, "An International Comparison of Sectoral Knowledge Bases: Persistence and integration in the pharmaceutical industry", Research Policy, 32(10).

Geuna A., 2000, "The Evolution of Specialisation: Public Research in the Chemical and Pharmaceutical Industries", Research Evaluation, 10: 67-79.

Landau R., 1998, "The Process of Innovation", in Arora A., Landau R. and Rosenberg N. (eds.), Chemicals and Long Term Economic Growth, John Wiley and Sons, New York.

Patel P. and Pavitt K., 1997, "The Technological Competencies of the World's Largest Firms: Complex and Path Dependent, but not Much Variety", Research Policy, 26, 141-56.

Sutton, J., 1991, Sunk Cost and Market Structure, Cambridge, MA: MIT Press.

Sutton, J., 1998, Technology and Market Structure, Cambridge, MA: MIT Press.

PART 1:

MARKET STRUCTURE AND INNOVATION

Chapter 1

MARKET STRUCTURE, COMPETITION, AND INNOVATION IN THE EUROPEAN AND US CHEMICAL INDUSTRIES.

Pedro L. Marín and Georges Siotis
Universidad Carlos III de Madrid-CEPR and Universidad Carlos III de Madrid

Abstract: This chapter applies the "bounds approach" to industry structure proposed by Sutton ((1991), (1998)). Part of the novelty in this exercise is that we work on the finest possible level of disaggregation. Also, we identify demand substitutability from direct industry sources. This allows us to carefully define markets, and identify R&D intensity for each of them. We then compare the relationship between market structure and innovation in Europe and the US. It emerges that both industries share many common features. In particular, it appears that the intensity of competition is similar across both side of the Atlantic. However, the European chemical industry is characterised by a lower degree of product homogeneity when compared to its US counterpart.

Key words: Bounds Approach; Market Structure; Innovation; Chemical Industry

1. INTRODUCTION

While the New Empirical Industrial Organization (NEIO) has provided testable hypotheses, many of the issues identified by "traditional empirical Industrial Organisation (IO)" have remained unanswered. The latter's focus has been on a cross-section of industries, rather than a single set of products (Schmalensee (1989)). One issue that has received particular attention is the relationship between market structure and Research and Development (R&D)

F. Cesaroni et al. (eds.), R&D, Innovation and Competitiveness in the European Chemical Industry, 25-43.

effort across industries. The existing empirical literature on this topic has provided mixed, if not contradictory, results (Cohen and Levin (1989)). Moreover, the Schumpeterian link between industry *concentration* and research effort has provided fuzzy results (Scherer (1990)).

Potentially, Sutton ((1991), (1998)) provides a unifying framework between these two traditions in empirical IO. On the one hand, his modelling approach makes extensive use of the game-theoretic tools that form the basis of theoretical developments in IO. On the other hand, his results are, under a set of reasonable assumptions, very general, as they encompass entire classes of models. This implies that the equilibrium concept does not include either a single or a reduced number of outcomes, but a wide variety of outcomes defined within a set of bounds.

Furthermore, one of the most interesting aspects of his approach is that his results point to testable hypotheses, both within industries, as well as across industries. For the purpose of this chapter, his contributions provide clear-cut predictions on the relationship between market size, the existence of alternative technological trajectories in R&D intensive industries, and concentration. In particular, Sutton predicts first, that as market size grows, industries with endogenous sunk costs, such as advertising and R&D expenditures, may not evolve in equilibrium towards fragmented market structures. Second, high R&D intensity industries' minimum concentration levels are affected by consumers' heterogeneity of preferences, since the latter fosters the existence of alternative technological trajectories. In that context, the heterogeneity in research trajectories yields more fragmented structures.

The drawback of this approach is that it is extremely demanding in terms of data. Consequently, existing empirical work have focused on industries where markets (or sub-markets) can be easily identified.[1]

Our results pertain to the chemical industry, which represents an important share of manufacturing output and employment. The interest of this industry is twofold. On the one hand, it encompasses different types of products in terms of the endogeneity of sunk costs, and a wide variety of consumer preferences. On the other hand, except for pharmaceuticals, advertising expenditures tend to be low since most products in the industry are not directed to final consumers. This allows us to clearly distinguish high from low R&D industries, and test the different predictions on these two types of industries. By combining distinct

[1] Lyons and Matraves (1996), Robinson and Chiang (1996), and Matraves (1999) perform empirical tests for several industries in the EU and the US that support Sutton's prediction about market size and market structure. Sutton (1998) himself tests his theory using four/five and seven/eight digit US Census data, but recognises that these data present some drawbacks.

data sources, we hope to obtain precise measures for the relevant variables. In particular, we work at a very low level of disaggregation, namely chemical products. By identifying alternative uses for these products, we are able to characterise the degree of demand substitutability. This allows us to construct markets that are the empirical counterparts of those presented in Sutton (1998).

The chapter is organised as follows: the next section reviews the theoretical literature and identifies a set of testable hypotheses. Section 3 describes the various data sets we make use of, and explains how the variables were constructed. Section 4 presents the results, while section 5 concludes.

2. THEORETICAL MOTIVATION

A brief review of the existing empirical evidence on the relationship between R&D intensity and concentration indicates that the results are at best mixed. Indeed, it seems that no solid empirical regularities can be detected in the data. On the theoretical side, we are faced with a set of models, each representing a specific situation, with each model providing a different, sometimes unique, equilibrium.

According to Sutton ((1991), (1998)), a single model may be adequate, but only for a narrow subset of R&D intensive industries. By contrast, the bounds approach does not require the identification of a unique equilibrium. Rather, it focuses on the whole set of outcomes that could form part of the equilibrium. This method proceeds by eliminating the set of realisations that cannot belong to any equilibrium given a sensible set of assumptions.

In order to characterise the set of possible outcomes, we need to define a stage game that represents firms' decisions. In the first period, that represents long run decisions, firms decide on entry and the level of sunk costs that they will incur (e.g., R&D expenditures related to a new product or process). The outcome may be a bundle of product characteristics, a set of locations, installed capacity, R&D or advertising expenditures. During the second period, that represents firms' short run decisions once entry has taken place and sunk costs have been paid, firms compete. Given any equilibrium configuration of products offered by different firms, there will be some corresponding set of equilibrium product prices, gross profit and industry sales. We thus need to identify which kind of configuration will be stable.

Clearly, there exist various ways to model each stage; the difficulty lies in the fact that there is no adequate discrimination process across models since the particulars of firms' decision process are not observable. Thus, the objective of the bounds approach is to reach conclusions independent of the form of competition and the entry process. Accordingly, Sutton (1998) defines an equilibrium configuration such that it satisfies two conditions: the "survival principle", which implies that firms are able to cover their fixed costs, and the "arbitrage principle", that refers to the fact that there are no empty spaces in equilibrium.[2]

This setting can be applied to exogenous and endogenous sunk cost industries. With respect to exogenous sunk cost industries, firms' decision at stage one is related to entry and the number of varieties to introduce in the market. In the context of a horizontally differentiated product market (exogenous sunk costs case), it is easy to show that as market size increases, so does the number of product varieties. Given that more competition (i.e. lower transport costs) implies lower price-cost margins, a larger market will be needed to attract entry since firms must pay their exogenous sunk costs with the profits that result from multiplying price-cost margins by sales. It is also the case that horizontal differentiation gives place to multiplicity of equilibria since one firm can produce several varieties of the product. Concentration increases in presence of scope economies, first mover advantages, and more homogeneity of consumers' preferences.

In the context of endogenous sunk costs, firms may spend on sunk costs (either advertising or R&D) in order to improve the characteristics of the products they sell. Sutton's framework crucially rests on two assumptions.[3] First, consumers' willingness to pay for a given product can be enhanced to some minimal degree by means of a proportionate increase in R&D or advertising outlays. Second, these outlays must be fixed (i.e. independent of the volume of output of the firm), have a small effect on unit costs, and be sunk (i.e. incurred at some earlier stage of the game and irrecoverable once the final stage is reached).

Notice that sunk costs are independent of market size, but profits derived from them are growing with market size, since firms following this policy set higher prices and obtain a larger market share. Thus, if firms face a large number of competitors, i.e., have a very small market share, then it will be profitable for some firm to outspend its rivals' sunk costs and capture a larger

[2] This concept is wider than the Sub-game Perfect Nash Equilibrium (SPNE), since the allocation of each firm in the strategy space does not need to be optimal given other players' allocation. A direct implication is that this approach encompasses the concept of SPNE.

[3] This argument applies in a context in which all consumers share the same ranking of preferences with respect to a specific characteristic of the product (quality).

market share. Hence, a configuration in which concentration lies below some critical value will be unstable.

When modelling R&D expenditures, two variables become central. R&D expenditures devoted to one group of products have limited spillovers on another group of products. Firms can develop many different technologies, each one related to a different group of products and must choose either to spend all their money on one trajectory or distribute it among several trajectories. This represents the choice between *escalation versus proliferation strategies*. Clearly, the choice between the two will depend on economies of scope and substitutability among product varieties. A high substitutability among varieties of the product implies that consumers are willing to buy the superior quality variety. Alternatively, if there are strong scope economies in R&D, innovations can be applied to all product varieties made by the firm. Therefore, we expect to observe higher concentration in industries with greater substitutability among product varieties and greater scope economies in R&D effort.

With respect to R&D intensive industries, it is possible to identify two variables related to the theory which are: the R&D/Sales ratio, and the degree of proliferation of distinct research trajectories. The intuition behind the first variable is the following. If R&D expenditures are ineffective in increasing consumers' willingness to pay for the product, then R&D intensity will be low. Consequently, we can deduce that industries with high R&D intensity can be characterised as endogenous sunk cost industries. The second variable is related to the degree of demand substitutability across groups of products and can be proxied by industry fragmentation into distinct product classes, i.e. the fraction of the industry's sales revenue accounted for by the largest single product class. We denote this homogeneity index h.

In R&D intensive industries, high substitutability will imply high concentration. If all firms have small market shares, escalation of R&D expenditures along a trajectory will be profitable (the high-spending firm will capture sales related to its own and other firms' trajectories). Moreover, firms will operate along a reduced number of trajectories (for each firm it is better to concentrate efforts along a single trajectory and build efforts on other people's results). Therefore, h will be high.

Low substitution will imply that in spite of the effectiveness of R&D expenditures, concentration may be low. This happens when there are many products associated to different R&D trajectories (low h). In this case, escalation does not lead to higher profitability, since market share is constrained to a single group of products. In such cases, a proliferation strategy is more effective in attaining a large market share.

Figures 1 and 2 below show lower bounds to concentration in exogenous (low R&D intensity) and endogenous (high R&D intensity) sunk cost industries depending on market size and substitutability. In low R&D intensive industries, the lower bound decreases with market size and the degree of substitutability is not expected to affect concentration. By contrast, in high R&D intensive industries, concentration is bounded away from zero and substitutability is negatively related to concentration.[4]

Figure 1. Lower bounds to concentration with respect to market size in exogenous and endogenous sunk cost industries.

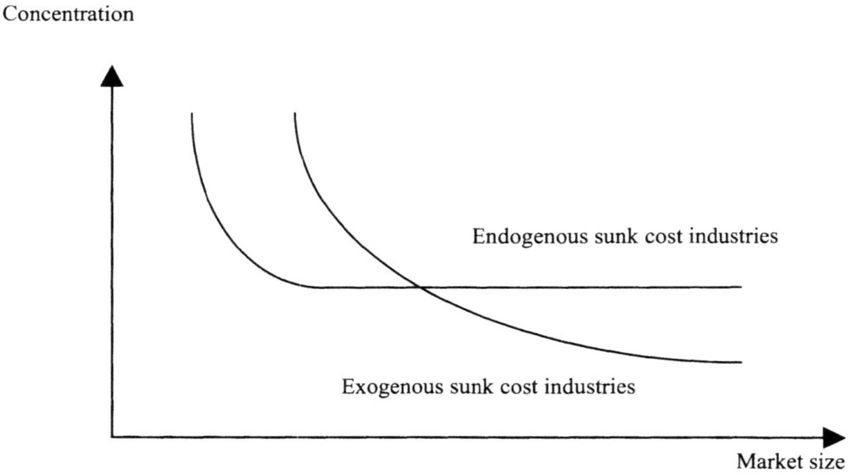

[4] As Sutton (1998) points out, one important aspect of this framework relates to the degree of aggregation. One could think that the whole matter could be solved by using the right degree of aggregation. Sutton provides a lengthy discussion on this problem.

Figure 2. Lower bounds to concentration with respect to the homogeneity index in high and low R&D intensity industries

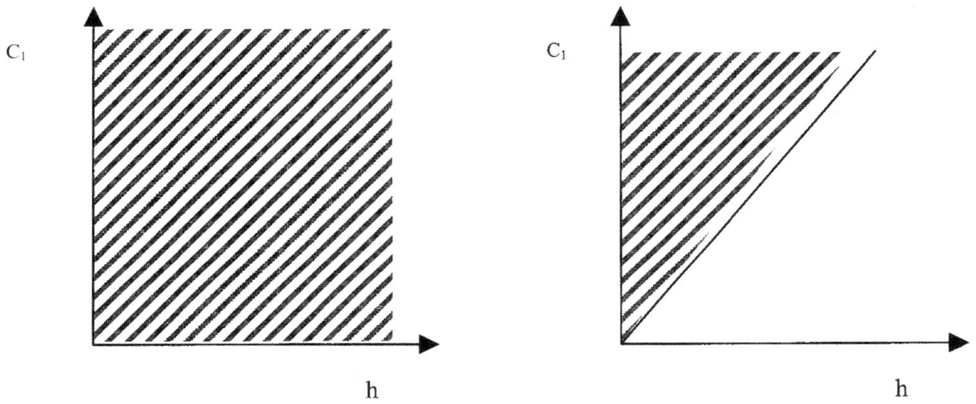

Note that the position of the bound in exogenous (low R&D intensity) sunk costs industries says something about the intensity of competition. A Northward displacement of that curve implies an increase in the intensity of competition.

3. DATA AND VARIABLES FROM THE CHEMICAL INDUSTRY

The chemical industry is formed by distinct markets where numerous firms are operating. Each market is characterised by a different R&D intensity and a different number of product classes, and presents a different structure. With the exception of pharmaceuticals, advertising expenditures are not particularly relevant since most of the products are sold as intermediate inputs. The remainder of the section proposes and applies a methodology based on technological substitutability to map chemical products into well defined markets.

Our data set has been constructed by combining information from various sources. The "central" database is Chemintell, which provides detailed information on 36,343 chemical plants worldwide. Right from the start, we have excluded pharmaceuticals from the analysis, given that this sector operates under very idiosyncratic conditions (see Appendices 8.1 and 8.2 in Sutton (1998) and Matraves (1999) for a discussion). Also, the information provided in Chemintell does not permit a proper treatment of that sector (for instance, there is no information on whether the drug produced is a brand product still under patent protection, or if it is a generic drug).

Chemintell contains data on plant capacity (installed or planned), measured in metric tons, for most of the plants in the database. In a reduced number of cases, whenever data on capacity was not reported (but other information was available), we assigned the average capacity obtained from other plants producing the same product. It also contains information on the location of plants, as well as the identity of their owners.

Of crucial importance for our purpose, the data pertains to chemical *products*. The database contains plants producing 2,279 different chemical products that are grouped into 14 broad categories such as petrochemicals, organic chemicals, inorganic chemicals, and the like. Clearly, this aggregation is of no use for our purpose. To properly test the theory, it is of uttermost importance to define markets adequately. Demand side substitutability is a key parameter at the time of constructing markets. While each product may embody distinct technologies, what matters is the degree to which a given product is perceived as a substitute for other products in the same market.

The empirical work that we are aware of has used a simple method to define markets by making use of existing official classifications. However, four/five digit groups are too broad and include many different markets. This is specifically true in the chemical industry. For instance, the four-digit group named "formulated pesticides" (CSO 2568), includes products such as fly paper (CSO 2568004), sheep dips (CSO 2568017), and plant hormones (CSO 2568014).[5]

We excluded many of the products found in Chemintell either because they represented a minute proportion of the sample (e.g., a product produced by only one plant in a non-OECD country), or because it proved impossible to obtain reliable information on their use. Thus, we examined, one by one, all possible outlets for the remaining 620 products. The latter account for 63.1% of the total number of plants. The relevant information was retrieved from the RISC database that provides a detailed input/output table for the chemical sector at a very low level of disaggregation. We were thus able to identify uses for 200 of our products. For the products that could not be classified using RISC, we relied on specialised publications, the Web, and trade journals. We were thus

[5] CSO stands for the classification used by the British Central Statistical Office.

able to distinguish between final products and intermediate inputs. Since we have to focus on final products only, we dropped all intermediate inputs. At the end of this process, we were able to identify 52 distinct groups of end-users, that is markets. We dropped 31 of them in order to err on the side of caution: end-users in these markets have access to substitutes outside the chemical industry (e.g.: textiles, cement and building, or automotive parts). While it may seem at first sight that we dropped many observations, this is not the case, as our 21 markets represent the bulk of final products in the chemical industry.[6] Our final sample consists of 363 products (15.9% of the initial sample) and 5554 plants located in the US and the EU.

In most cases, a given product was identified as having more than one possible use. Consequently, we had to decide how to allocate recorded capacity for each product across our markets. We decided on a simple rule, namely to split capacity evenly among the markets in which the product appears. This avoids ad-hoc judgements that would have been necessary in the event of choosing an alternative method.[7] With respect to geographical markets, we considered two regions: the United States, and "Europe". The latter is made-up of France, Germany, the UK, Italy and Spain.

With this information we constructed the homogeneity index, h, as the percentage of installed capacity accounted for by the most popular product within each market. Before constructing the variables measuring concentration, we had to aggregate plant level information into firm level data. With this information in hand, we were able to compute measures of concentration, and in particular the four-firm (C_4) concentration ratio.[8]

The last step pertains to the partitioning of markets according to R&D intensity. Unfortunately, Chemintell does not contain data on R&D expenditures. To alleviate this problem, we made use of Worldscope, a database that provides information on the 1,500 largest R&D spenders in Europe. We retrieved the R&D of all the firms that appear in Worldscope and that are active in one (or more) of our 21 markets. For each of our markets, we computed the average R&D intensity using the data from these firms. We calculated the weighted average, using sales as weights. In that exercise, the cut-

[6] Recall that we have excluded pharmaceuticals right from the start.

[7] An alternative is to assume that a 100% of capacity for a given product can be allocated to each of the markets in which it appears. We tried that route, and the empirical results were qualitatively similar. We ended-up choosing the alternative proposed in the text because we feel it is more adequate.

[8] The results are robust to alternative measures of concentration such as the one-firm (C_1) concentration ratio.

off value we chose for R&D intensity is 2%.[9] As can be seen in Table 1, this is a natural cut-off value in the sense that it splits the sample in two clearly defined groups in terms of R&D intensity.[10]

[9] 2% may appear as low at first sight; it should however be remembered that Worldscope provides R&D intensity for the entire firm. In the chemical sector, these are typically large multiproduct firms, which tends to depress measured R&D intensity.

[10] Table 1 only reports R&D intensity for 19 of our 21 markets. The reason is that for two markets (batteries and disinfectants), we were not able to obtain R&D expenditures from Worldscope. Thus, these two sectors were classified as low R&D, since large R&D spenders (the firms found in Worldscope) were not active in these markets.

Table 1. Data on R&D intensity retrieved from Worldscope

Markets	SIC code	Description	No. of firms	Weighted R&D to sales ratio[1]	Max. R&D to sales ratio[2]	R&D/Sales of the largest firm[3]
Adhesives	2891	Adhesive & sealants	3	2.95	4.07	3.33
Agrochemicals	2879	Pesticides & agric chem., n.e.s.a	3	2.77	8.94	0.95
Alloys	2821	Plastic materials, synthetic resins, and nonvulcanizable elastomers manuf.	7	4.47	5.71	4.59
Coatings	2952	Asphalt felts & coatings m	1	0.50	0.74	0.50
Dyes	2816	Inorganic pigments	1	1.56	2.09	1.56
	2865	Cyclic organic crudes, interm., organic dyes, & pigments	8	6.75	7.66	6.96
	Total		9	6.53	7.66	6.96
Fertilizers4 (exc. nitrogenous)	2875	Fertilizers, mixing only	1	0.51	0.58	0.51
Fibres	2823	Cellulosic manmade fibers	3	3.10	4.89	3.48
	2824	Manmade organic fibers, exc. cellulosic	2	1.50	2.27	1.58
	Total		5	2.58	4.89	3.48
Film	2671	Packaging paper and plastics film, coated and laminated manuf.	4	1.91	2.87	2.01
	3081	Unsupported plastics film & sheet	1	0.07	0.12	0.07
	Total		5	1.82	2.87	2.01
Fuel	1221	Bituminous coal & lignite surface mining	1	2.42	2.42	2.42
	1222	Bituminous coal underground mining	2	0.76	2.23	0.67
	1311	Crude petroleum & natural gas	9	0.83	5.15	0.75
	1321	Natural gas liquids	2	0.41	0.63	0.41
	1381	Drilling oil & gas wells	1	2.19	2.19	2.19
	1382	Oil & gas field exploration services	2	0.97	1.97	0.70
	1389	Oil & gas field services, n.e.s.	2	1.37	3.15	0.56
	Total		19	0.76	5.15	0.67
Gas handling	1311	Crude petroleum & natural gas	9	0.83	5.15	0.75
	1321	Natural gas liquids	1	0.41	0.63	0.41
	2813	Industrial gases manuf.	4	2.24	5.33	1.52
	Total		14	1.76	5.33	0.75

Table 1 (cont.). Data on R&D intensity retrieved from Worldscope

Markets	SIC code	Description	No. of firms	Weighted R&D to sales ratio[1]	Max. R&D to sales ratio[2]	R&D/Sales of the largest firm[3]
Lubricants	2992	Lubricating oils & greases	3	1.93	2.39	2.12
Nitrogenous fertilizers	2873	Nitrogenous fertilizers	3	1.34	2.08	1.03
Oil & gas pipes	3084	Plastics pipe manuf.	1	1.23	1.32	1.23
	3494	Valves & pipe fittings, n.e.s	2	0.23	0.26	0.26
	3498	Fabricated pipe & pipe fittings	3	2.39	3.97	2.06
	Total			1.33	3.97	0.26
Paints	2851	Paints, varnishes, lacquers, enamels & allied products	8	1.87	5.34	1.90
Rubber	2822	Synthetic rubber	1	1.17	1.91	1.17
	3021	Rubber & plastic footwear	2	2.59	2.59	2.59
	3052	Rubber, & plastic hoses & belting	2	3.49	4.28	1.05
	3061	Molded, extruded & lathe-cut rubber	1	1.73	2.19	1.73
	3069	Fabricated rubber n.e.s.	4	2.55	3.68	3.37
	Total		10	2.43	4.28	3.37
Soap	2841	Soap & other detergents, exc. specialty cleaners	1	1.33	1.44	1.33
	2842	Specialty cleaning, polishing & sanitation prep. manuf.	2	1.37	1.77	1.37
	Total		3	1.34	1.77	1.33
Surfactants	2842	Specialty cleaning, polishing & sanitation prep. manuf.	2	1.37	1.77	1.37
	2843	Surface active agents, finishing agents, sulfonated oils & assistants	2	1.21	1.79	1.21
	Total		4	1.23	1.79	1.21

Source: IPC, Worldscope and Chemintell databases, and own calculations
[a] n.e.s.: Not elsewhere specified
[1] Weighted average for all firms during the six years of data.
[2] Maximum reported by firms during the six years
[3] Weighted average for the largest firms during the six years of data, with size measured as sales during the last year of data.
[4] This market groups Potassium, Phosphorous, and Sulphur fertilizers

In most cases, it proved straightforward to determine the R&D associated with each product. However, in some instances, markets were served by both high and low R&D intensity firms. This arose quite often when we were using information from multiproduct firms that were simultaneously active in pharmaceuticals. These firms often hold a wide product portfolio whose R&D intensities differ substantially (e.g. acids and pharmaceuticals).

To alleviate possible problems linked with this indeterminacy, we used an additional database on patents that contains information on 201,791 patents from the chemical sector. Using the fine disaggregation of the International Patent Classification (IPC), we were able to match the number of patents to some of our products. Overall, the markets that were identified as high R&D using the information contained in Table 2 also turned to markets where the number of patents is large (with "large" being defined as more than 500 patents).[11] The only exception is the market "Fuel", for which we identified 7179 patents but whose R&D intensity is 0.76%.[12] This market was thus deemed to be high R&D, despite the fact that it does not fulfil the 2% criteria.[13]

Finally, we had to construct a variable that measures the extent of the market. To this end, we proxied market size (denoted as *S*) by the sum of reported capacity of all the products that form our markets. We then computed the size of the median plant in each of our markets, which yields an approximation of the minimum efficient size of operation (denoted by *MES*), that is, of exogenous sunk costs. Thus, the variable that measures the extent of the market is defined as the natural logarithm of S over *MES*, i.e. ln(S/*MES*).[14]

[11] The information on patents also confirmed that batteries and disinfectants were low R&D, as the number of patents turned out to be very small (see also footnote 11).

[12] In this market, large R&D expenditures are dwarfed by the volume of sales.

[13] For practical purposes, wrongly classifying a sector as high R&D while it is not can only make the empirical analysis more arduous, while the converse is not true (the theory makes more restrictive predictions regarding high R&D sectors).

[14] The use of this variable may involve a potential bias since concentrated industries are characterised by large firms that operate large plants (see Davies (1980)). However, there is no evidence of such a correlation for the full sample, or across industries at the country level, or across countries at the industry level.

4. EMPIRICAL FINDINGS

The frontiers depicted in Figures 3.a and 3.b present the relationship between (a transformed measure of) concentration and the extent of the market. Following Sutton, the variable C_4^* is defined as:

$$C_4^* = \ln \frac{C_4}{1 - C_4} \qquad (1)$$

The reason for applying a logit transformation is that C_4 is characterised by an upper limit (C_4=1), and thus the residuals are not identical at all values of the independent variable ln(*S/MES*) (defined below).

The next step consist in choosing a schedule to describe the bound. The fitted schedule that describes the lower bound is the one proposed by Sutton (1991), that is:[15]

$$C_4^* = d + \frac{b}{\ln(S / MES)} \qquad (2)$$

In Figure 3.a, we have a plot relating the extent of the market to the degree of concentration in low R&D industries (that is, exogenous sunk costs industries). As can be readily seen, the empirical results are in line with the theoretical predictions of Sutton (1991). This holds both for Europe and the US: industry concentration falls as market size increases.

While no clear cut predictions exists as to whether the European and US sectors should be clustered, it turns to be so. That is, for a given industry, the EU-US plots stand close by. The correlation between concentration in Europe and the US stands at 0.79. As a general rule, and perhaps surprisingly, concentration is higher in the EU in low R&D intensity industries.[16] In ten sectors (out of a total of fourteen), concentration is higher in the EU. The exceptions are fertilizers (phosphorous and potassium), film, and paints, with the difference being small in all four cases. Overall, this indicates that the degree of competition is slightly higher in the EU in these low R&D intensity industries.

[15] Apart from the fact that is appears reasonable, Sutton does not provide further explanations for this choice of schedule.

[16] This may be due to the fact that our "definition" of Europe only includes the five largest EU economies.

Figure 3a. Low R&D industries

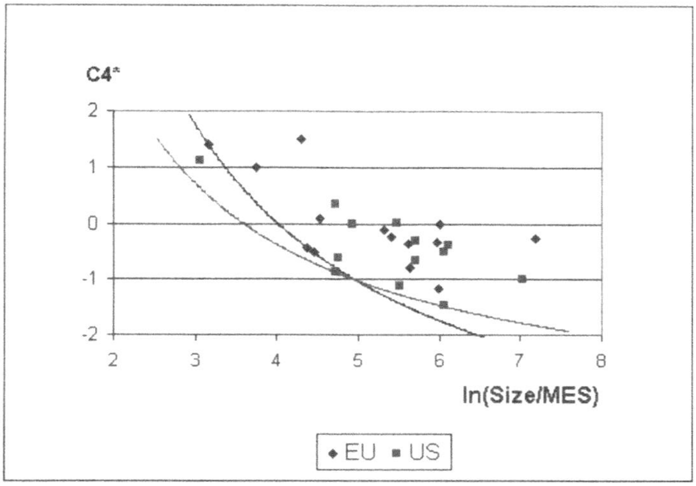

Figure 3b. High R&D industries

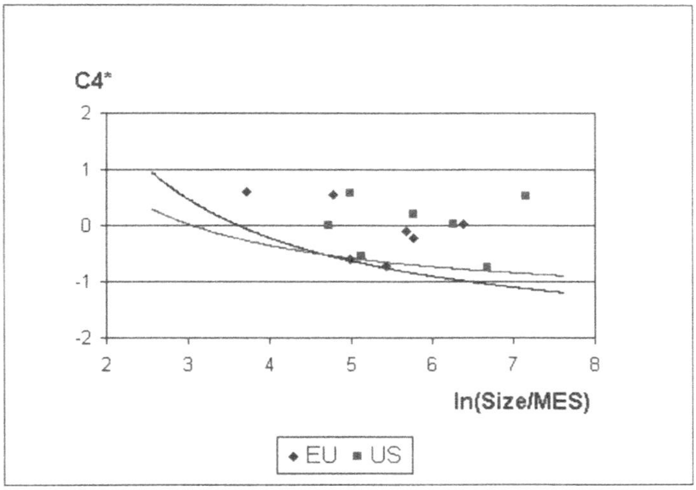

Figure 3.b depicts the same plot for R&D intensive industries. Again, the shape of the bounds confirms the theoretical predictions: concentration initially declines with market size, but stops doing so once a threshold is reached. The EU-US pairs are somehow clustered together, and concentration on both sides of the Atlantic is weakly correlated across industries (coefficient of correlation of 0.28). Of the seven sectors, concentration is marginally higher in the US in four cases, and significantly so in one case. Perhaps surprisingly, concentration is higher in Europe in dyes and adhesives. For these two industries we may expect an increase in concentration in the US. By the same token, we may expect an increase in concentration in Europe in the remaining five sectors.

The fitted schedules also provide information on the lower bound to concentration as market size tends to infinity. As $S/MES \to \infty$, the constant d from equation (2) yields the minimum level of concentration. These values are given in Table 2 below.

Table 2 . Estimation of lower bounds for market concentration as market size grows

R&D intensity	Market	$d = C^{*}_{4,\infty}$	b	$C_{4,\infty}$
High	EU	-2.14	7.71	10.52
	US	-1.39	4.16	19.94
Low	EU	-5.22	2.87	0.54
	US	-3.61	12.96	2.64

The results are satisfactory both for Europe and the US, as asymptotic concentration is clearly bounded away from zero in high R&D intensity industries. Asymptotic concentration is clearly higher in the US compared Europe, both in low and high R&D intensive industries. A possible explanation for this finding is that the US market has been fully integrated for a much longer period.

Figure 4.a. and 4.b relate concentration to our homogeneity index h in low and high R&D intensity industries respectively. The plots are consistent with Sutton's (1998) theoretical predictions (as depicted in Figure 2). We observe a random scatter for low R&D intensity industries (Figure 4.a), while the South-East corner of Figure 4.b is empty. Recall that the h parameter measures the degree of substituability on the demand side. In low R&D intensity markets, the homogeneity index is higher in the US in eight out of fourteen cases. In high R&D markets, homogeneity is systematically higher in the US. Overall, this suggests that consumer preferences are less homogeneous in Europe. This is perfectly consistent with the fact that market integration has been a more recent phenomenon in the Old Continent compared to the US. Also, industry production structures display strong inertia. This implies that some degree of fragmentation would persist even following full market integration. This may also indicate that the EU chemical industry is more focused on niche markets characterised by distinct technological trajectories.

The case of dyes and adhesives, both of which are high R&D markets, deserve a final comment. In the latter cases, concentration is higher in Europe but the homogeneity index is lower. This is consistent with the existence of distinct technological trajectories that display economies of scope among themselves. Thus, EU firms in these markets seem to have successfully exploited these opportunities.

Figure 4a. Low R&D industries

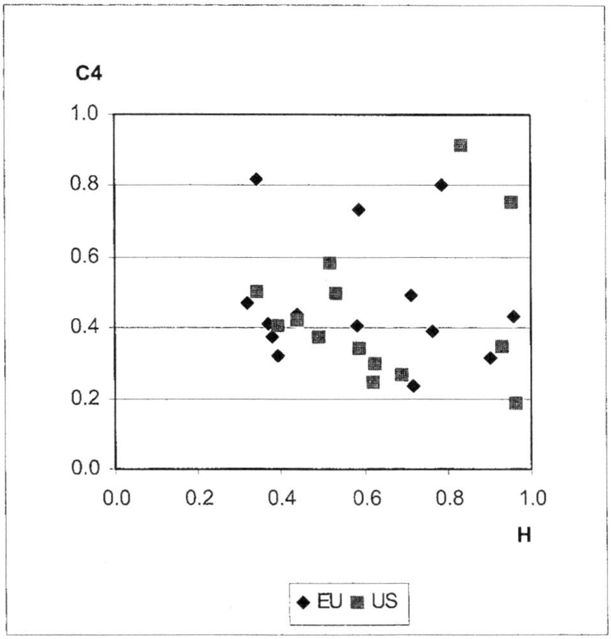

Figure 4b. High R&D industries

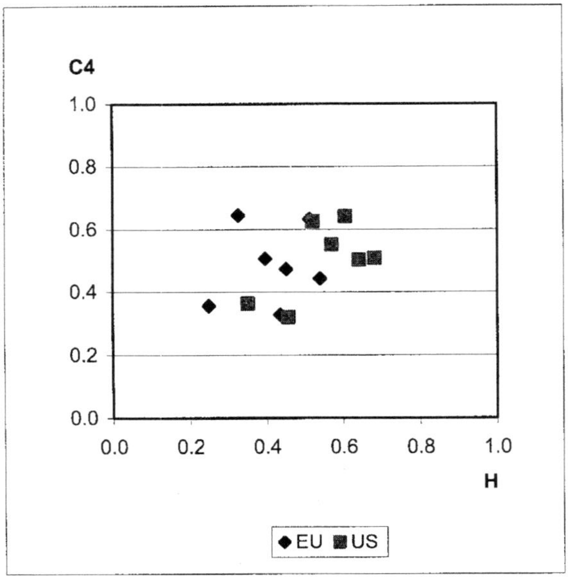

5. CONCLUSIONS

The evidence presented in this chapter is two-fold. First, we provide evidence supporting Sutton's (1991, 1998) theoretical predictions. In a companion paper, we report additional statistical evidence that confirms these findings (see Marín and Siotis (2002) for an in depth technical analysis). Second, we have compared industry structures and the intensity of competition on both sides of the Atlantic. Overall, our findings indicate that industry structures are quite similar. Also, there is weak evidence suggesting that, in at least some low R&D industries, competitive pressure is stronger in the EU compared to the US. However, the reverse is also true: in some industries, competitive conditions are sharper in the US, particularly in R&D intensive ones. We also find that consumers preferences are systematically more homogeneous in the US compared to the EU, a finding in line with priors. The similarity between the US and European chemical industries in terms of market structure and competition point in the same direction as the findings pertaining to Mergers and Acquisitions reported in chapter 2 of this book.

6. REFERENCES

Cohen, W. and R. Levin, (1989), "Empirical Studies of Innovation and Market Structure", in Schmalensee, R, Willig, R., (eds.) Handbook of Industrial Organization, Vol. II, pp. 1060-1098.

Davies, S., (1980), "Minimum Efficient Size and Seller Concentration: An Empirical Problem", The Journal of Industrial Economics, 29, pp.287-302.

Lyons, B. and C. Matraves, (1996), "Industrial Concentration" in Davies, S, and Lyons, B, et al, Industrial Organization in the European Union: Structure, Strategy and the Competitive Mechanism, Oxford University Press, Oxford.

Marín, P., and Siotis, G., (2002) "Innovation and Market Structure: An Empirical Evaluation of the "Bounds Approach" in the Chemical Industry", CEPR DP #3162, January.

Matraves, C., (1999), "Market Structure, R&D and Advertising in the Pharmaceutical Industry", The Journal of Industrial Economics, 47 (2), pp.169-194.

Robinson, W. and J. Chiang, (1996), "Are Sutton's Predictions Robust? Empirical Insights into Advertising, R&D, and Concentration", The Journal of Industrial Economics, 44 (4), pp.389-408.

Scherer, F. M. and D. Ross, (1990), Industrial Market Structure and Economic Performance, Third Edition, pp. 644-660, Houghton Mifflin Company, Boston.

Schmalensee, R. T., (1989) "Inter-industry Studies of Structure and Performance" Schmalensee, R, Willig, R, (eds.) Handbook of Industrial Organization, Vol. II, pp. 951-1009.

Sutton, J., (1991), Sunk Cost and Market Structure, Cambridge, MA: MIT Press.

Sutton, J., (1998) Technology and Market Structure, Cambridge, MA: MIT Press.

Shreve and Brink (1977), Chemical Process Industries, McGraw Hill.

Chapter 2

MERGERS AND ACQUISITIONS IN THE CHEMICAL INDUSTRY: SIMILARITIES AND DISIMILARITIES ACROSS THE ATLANTIC

Carmine Ornaghi*, Pedro L. Marín** and Georges Siotis*
*Universidad Carlos III de Madrid and **Universidad Carlos III de Madrid-CEPR

Abstract: This chapter provides a quantitative description of mergers and acquisitions (M&As) in chemicals and related industries during the period 1985-1997, with the aim of identifying the main features characterising the globalisation of the European chemical industry. The analysis is based on data drawn from the "SDC International Mergers & Acquisitions" database. We compare US and European firms' M&A strategies in terms of targets' location, sector, and mode of acquisition (minority vs. majority). Our main findings is tha M&A strategies are strinkingly similar on both sides of the Atlantic, and that olipolistic rivalry, rather than comparative advantage, drives M&A flows in the chemical sector. Last, we do find some evidence that, in R&D intensive sectors, European chemical firms acquire US based assets in order to maintain and expand their knowledge base.

Key words: Mergers and Acquisitions; Multinational Firms; Investment flows; Chemical Industry

1. INTRODUCTION

The chemical industry operates on a global scale. The internationalisation of the sector can be traced back many decades, and chemical firms are among the world's largest. Most of them are highly diversified, both in terms of product portfolio as well as geographic span. Every major player in this industry operates plants in almost any significant geographical market.

F. Cesaroni et al. (eds.), R&D, Innovation and Competitiveness in the European Chemical Industry, 45-68.

While Japan has built a substantial chemical industry since World War II and some emerging market economies have grown on the World stage, the industry remains dominated by North American and European groups (Kang and Johansson (2000)). International expansion has been fuelled by large Foreign Direct Investment (FDI) flows, a substantial proportion of which took the form of Mergers and Acquisitions (M&As).

The explanations put forward to account for the observed patterns of foreign investment are drawn from trade theory, industrial organisation, and the international business literature. Dunning's (1981) eclectic paradigm attempts to bring these different strands of the literature together, and it remains a standard reference for scholars studying FDI. His approach, typically referred to as the Ownership-Localisation-Internalisation (OLI) paradigm, stresses that three factors must be present for FDI to occur.

First, the potential or actual multinational must be endowed with some firm-specific ownership advantage not available to host country firms. This requirement is usually put forward to explain why a multinational firm, which is a priori at a disadvantage compared to local firms in the host country, can profitably invest abroad. The original idea is due to Hymer, and it is commonly accepted as a necessary condition for FDI to occur. However, this proposition has recently been challenged by Fosfuri and Motta (1999) who show that in an oligopolistic context, FDI can occur even in the absence of any firm specific ownership advantage. In the latter paper, oligopolistic rivalry drives firms abroad.

Second, there must be localisation advantages associated with foreign production. They range from the absence of exporting costs, such as tariffs and transport costs, better market access, or the availability of cheaper inputs such as labour and raw materials.

Third, internalisation advantages determine the choice of FDI as opposed to arm's length market transactions, such as licensing agreements, which may entail the diffusion of the multinational's assets to actual or potential competitors (see Markusen (1995) for a discussion). Thus, internalisation is often viewed as a device to limit the diffusion of firm specific intangible assets. However, while internalisation may be a superior mechanism compared with an arm's length transaction, it is unlikely that it will prevent diffusion altogether. This is particularly relevant in the context of FDI, since empirical evidence suggests that multinationals are found in industries and market segments in which the presence of intangibles is pervasive.

Several recent papers have suggested that access to new technology is an increasingly important motivation for firms locating their activity abroad. Serapio and Dalton (1999) argue that much of the globalisation of innovative activity has involved foreign firms locating their activities in the US in order to source leading edge technology.

Chemical firms have played a prominent role in the wave of international M&As, accounting for some of the biggest international mergers of the last decade. Recent examples include British Petroleum-Amoco and Exxon-Mobil in the petroleum industry and Astra-Zeneca Group and Pfizer-Warner Lambert in pharmaceuticals. Among the top ten industries for cross-border M&As in 1998, “Oil and Gas Extraction” and “Chemical and Pharmaceuticals” industries rank 1st and 7th, respectively (Kang and Johansson (2000)).

The aim of this chapter is to conduct an exhaustive quantitative description of mergers and acquisitions (M&As) in chemicals and related industries. Using insights from the conceptual framework presented above, we analyse M&As during period 1985 to 1997, with the aim of identifying the main features characterising the globalisation of the European chemical industry. In particular, we concentrate on similarities and dissimilarities across the Atlantic with respect to domestic (intra-regional) and international M&As. This exercise is carried out by focusing on the type of industries involved, percentage of ownership acquired, and modes of entry.

This chapter is organised as follows. Section 2 describes the dataset, while section 3 analyses the origin of acquiring firms and target firm’s location. In section 4, we attempt to identify sector differences in the pattern of M&As in the US and Western Europe. In section 5, we examine the diversification strategy and the extent of ownership control exercised by parents on subsidiaries. In particular, we assess whether minority acquisitions are more likely when firms enter new markets. Section 6 concludes.

2. THE DATA

The central database used is named the "SDC International Mergers & Acquisitions Database". This database records all large M&As since the late seventies for US companies, while all major international transactions are recorded as of 1985. The last recorded transactions are dated at the beginning of 1998. Since the SDC database originates in the US, it is probably biased towards recording transactions involving US entities, and possibly European ones. This caveat must be borne in mind when absolute comparisons are made. However, this should not affect relative comparisons, that is when shares and growth rates are computed.

We have extracted transactions where the primary SIC code of either the acquirer or the target firm belongs to one of the following industries: chemicals and allied products (SIC 28), petroleum refining (SIC 29), rubber and plastic (SIC 30) or oil and gas extraction (SIC 13).

The information provided for each transaction is both qualitative and quantitative. It includes variables such as the date of transaction, the 4-digit affiliation of the companies involved, name and nationality of the target firm, of the acquirer and of the ultimate parent company, the percentage shares acquired, and the form of the transaction. In addition, there are some financial data about the target firm such as sales, intangible assets, and R&D expenditures.

The information we retrieved are the date of the transaction, the name and nationality of the target firm and of the ultimate parent, the SIC codes of the firms involved, the percentage of shares acquired, and R&D expenditures. Flows of investment are described in terms of number of transactions reported in the database. This is obviously an imperfect measure of FDI, as it does not take into account the cost of the acquisition and the turnover of the target company. Nevertheless, it is possible to describe the changes in relative flows of investments, and how the presence of European and American firms has changed over time in the main regions of the world.

We define seven regions of investment:

i) Asia & Oceania, except Japan and the Middle-East;
ii) Japan;
iii) Middle East and North Africa (Maghreb and Mashrek countries);
iv) Eastern Europe;
v) Western Europe;
vi) North America: United States and Canada;
vii) Rest of the World: Latin-America and Africa (except North-Africa).

A transaction will be called intra-regional if the acquirer and target firms belong to the same region (e.g. Western Europe), and international or cross-border if the firms involved originate in two distinct regions.

As mentioned above, the data set reports the primary SIC of both the target and the ultimate parent at 4-digit level. While the primary SIC of the target is just a single number, the SIC of the ultimate parent range from a minimum of one to a maximum of 29 entries, with an average of 5 different 4-digit codes. By making use of this data, we have identified different diversification strategies according to whether the subsidiary's product line corresponds to one of the acquirer's product lines. For example, whenever the ultimate acquirer and the target company have the same 2-digit SIC code, we categorise the transaction as a horizontal M&A. A complete and detailed explanation of all the diversification strategies identified is presented in Section 5.

Data on the equity acquired is only available for 62% of transactions. For the latter, we have divided the transactions according to the extent of ownership control in two different groups that we label as majority (share greater or equal to 50%) and minority owned (share less than 50% of the equity). In section 5, we analyse diversification strategies and the extent of ownership control exercised on subsidiaries for different regions and industries. In addition, we identify industries with higher technological contents in order to assess whether the entry mode of the acquirers depends on the R&D intensity of the target firm's industry.

3. OVERVIEW OF M&AS FLOWS

Table 1 presents the characteristics of M&As for the period 1985-1997. Out of 25,904 transactions listed, 14,007 target firms are located in North America and 7,293 in Western Europe, representing 82% of the transactions. Moreover, European and North American parent companies account for more than 80% of the transactions listed. This confirms the widespread finding that industrialised countries are simultaneously the main source and destination of FDI. Quite surprisingly, Japanese acquirers account for only 2% of the total number of acquisitions and this percentage is smaller than 1% if we consider the location of the target firm. This has to be related with the data reported in Arora, Gambardella and Garcia Fontes (1998) that show that the percentage of chemical plants owned by Japanese companies is 12.5% while the percentage of plants located in Japan is about 12.2%. The combination of these findings confirm the well known preference of Japanese firms for greenfield FDI. It also confirms the dominance of European and US players on the global M&As map.

The bottom part of the table reports the distribution of mergers and acquisitions by year of transaction. The number of operations has increased over time, from 3,509 cases for the period 1985-1988 to 9,197 for the period 1995-1997.

The increase in the number of transactions is confirmed for most of the regions defined above. Allocating the number of M&As by region of target firm and year of transaction, we detect two main trends. First, M&As involving Eastern European firms have increased notably during the period 1995-1997, which corresponds to the introduction of market oriented reforms in these economies. Second, the evolution of M&As in Western Europe does not show the steady increase that characterises all the remaining areas. Table 2 shows that the percentage of intra-regional transactions in that region during the period 1995-1997 (28.3%) is smaller than the one of the previous period (31.7%). On the other hand, international acquisitions of American firms undertaken by European acquirers are sensibly higher in the last three years.

Table 1. Mergers and Acquisitions in Chemicals 1985-1997

Region of the Target Company	Number	%
Asia	2134	8.24
Japan	225	0.87
Middle East and North Africa	301	1.16
Eastern Europe	762	2.94
Western Europe	7293	28.15
North America	14007	54.07
Rest of the World	1079	4.17
Not disclosed	103	0.40
Total	25904	100
Region of the Ultimate Acquirer	Number	%
Asia	1548	5.98
Japan	559	2.16
Middle East and North Africa	232	0.90
Eastern Europe	296	1.14
Western Europe	7784	30.05
North America	13256	51.17
Rest of the World	510	1.97
Not disclosed	1719	6.64
Total	25904	100
Periods when the M&A took place	Number	%
1985-1988	3509	13.55
1989-1991	6036	23.30
1992-1994	7162	27.65
1995-1997	9197	35.50
Total	25904	100

Source: SDC International M&A Database.

Table 2 also indicates that the total number of cross border M&As in Western Europe is greater than that of North American acquirers (1,507 against

1,165). Combined with the findings reported in Table 1, this indicates that European M&A activity is more outward oriented. Moreover, the distribution of the transactions over the four periods in these two regions is different. While international M&As from North America to Europe increases over time, those in the opposite direction follow a cyclical pattern. Figures in the bottom part of the table suggest that when European acquirers conclude a larger number of international acquisitions, they tend to reduce intra-regional investments. This is consistent with the idea that there is a "pool" of investments whose destination is influenced by international factors, which provides direct evidence that the industry operates on a global scale.

Table 2. Mergers and Acquisitions across the Atlantic

		Target's Region			
Acquirer's		North America		Western Europe	
Region	Period	Number	%	Number	%
North America	1985-1988	1793	15.99	165	14.16
	1989-1991	2484	22.15	252	21.63
	1992-1994	2897	25.84	368	31.59
	1995-1997	4038	36.01	380	32.62
	Total	**11212**	**100**	**1165**	**100**
Western Europe	1985-1988	357	23.69	663	12.45
	1989-1991	420	27.87	1466	27.53
	1992-1994	302	20.04	1688	31.70
	1995-1997	428	28.40	1508	28.32
	Total	**1507**	**100**	**5325**	**100**

Source: SDC International M&A Database.

Table 3 shows the total number of intra-regional M&As. With few exceptions, most of the transactions have been concluded between companies belonging to the same area as highlighted by the grey-shaded figures. Although international M&As are important, domestic firms are still the main investors in each of the seven regions. So much for the fears of massive delocalisation triggered by globalisation.

If we focus on international transactions (that is, where the region of the acquirer does not coincide with the region of the target firm) the first feature that emerges is that North America plays a major role in attracting inward M&As accounting for 36.8% of the transactions, against 27.6% for Europe. By contrast, Europe accounts for a higher percentage of outward investments, with

43.7% of transactions, against 36.7% for North America. This reflects broader FDI trends world-wide that have seen the US becoming a net recipient of FDI, while Europe has been characterised by net outflows. Quite interestingly, while during the period 1992-1994 the number of M&As from US to Europe are higher than in the opposite direction, the following three years record an inversion of this trend. As we indicate in the next section, this change is likely to be driven by investments in high-tech sectors such as Pharmaceutical, Biotech, and Research and Engineering.

Table 3. Geographical Distribution of M&As

	Target's Region							
	Asia	Japan	Middle East	Eastern Europe	Western Europe	North America	Rest of World	Total
Acquirer's Region	Number (%)	Number (%)	Number (%)	Number (%)	Number (%)	Number (%)	Number (%)	Number (%)
Asia	1240 (62.1)	5 (2.3)	18 (6.8)	11 (1.9)	94 (1.4)	156 (1.2)	22 (2.3)	**1546** (6.4)
Japan	69 (3.5)	125 (56.6)	7 (2.6)	1 (0.2)	107 (1.6)	236 (1.8)	11 (1.1)	**556** (2.3)
Middle East	9 (0.5)	2 (0.9)	103 (38.9)	12 (2.1)	72 (1.1)	25 (0.2)	8 (0.8)	**231** (1.0)
Eastern Europe	14 (0.7)	0 (0.0)	2 (0.8)	253 (44.0)	20 (0.3)	3 (0.0)	3 (0.3)	**295** (1.2)
Western Europe	359 (18.0)	50 (22.6)	67 (25.3)	197 (34.3)	5325 (77.9)	1507 (11.4)	223 (23.2)	**7728** (32.1)
North America	299 (15.0)	39 (17.6)	65 (24.5)	93 (16.2)	1165 (17.0)	11212 (84.7)	354 (36.9)	**13227** (54.9)
Rest of World	7 (0.4)	0 (0.0)	3 (1.1)	8 (1.4)	56 (0.8)	94 (0.7)	339 (35.3)	**507** (2.1)
Total	**1997** (8.3)	**221** (0.9)	**265** (1.1)	**575** (2.4)	**6839** (28.4)	**13233** (54.9)	**960** (4.0)	**24090** (100)

Source: SDC International M&A Database.

Notes:
The percentage in each column is w.r.t. the total of the column (e.g., 2.1=1240/1997).
The percentage in the bottom line is w.r.t. the sum of all the totals (e.g., 8.3=1997/24090).
The cells in grey represent intra-regional investments.

Finally, Table 4 provides further details on investment flows within Europe. A relevant feature is the position of United Kingdom, both as a source and destination of investments. About 40% of the acquisitions have been targeted at British firms, and 45% of transactions have involved British acquirers.

However, Table 4 indicates that most of the transactions are between companies belonging to the same country: domestic investments go from a minimum of 32% for Spain to a maximum of 86% for the United Kingdom. When we concentrate on cross-border M&As, figures are quite different. Although United Kingdom still plays a dominant role in outward M&As followed by France and Germany, it ranks only third in attracting inward investment. This is somewhat surprising, as the UK is often portrayed as the most FDI-friendly country in the European Union (EU).

Summing up, there are four main patterns that emerge from this analysis. First, while the number of both intra-regional and international deals by American acquirers have constantly increased over the period, there is a slowdown in the number of European intra-regional M&As during the last three years of the sample. Second, European and North American companies are the main investors in all the regions, while Japanese companies seem to play a secondary role in international M&As. Third, domestic investment in industrialised countries is considerably higher than foreign investment. Indeed, European and American target firms are mainly acquired by European and American firms, respectively. This finding is confirmed by an analysis at the regional level for Western Europe. Still, this finding cannot be taken as evidence against the globalisation of the chemical sector. Since international M&As are typically riskier than domestic ones, it is fairly natural that the absolute value of intra-regional transactions outweigh the international ones. Moreover, Table 2 shows faster growth for international transactions compared to intra-regional ones, confirming that the industry operates on an global scale. Finally, there is a higher flow of investments from Europe to North America than vice versa. This may be considered as mild evidence indicating the existence of "technology sourcing", since, on average, US firms are more likely to be operating at the technological frontier. We come back to this point in the next section.

Table 4. Distribution of M&As between Western European Countries

	Target's Region							
	France	Italy	Holland	Spain	Switzer-land	U.K.	Germa-ny	Total
Acquirer's Region	Number (%)	Number (%)	Number (%)	Number (%)	Number (%)	Number (%)	Number (%)	Number (%)
France	253 (61.4)	36 (14.2)	11 (5.6)	36 (21.6)	4 (5.7)	52 (4.6)	64 (11.9)	**456** (16.6)
Italy	6 (1.5)	113 (44.7)	1 (0.5)	8 (4.8)	3 (4.3)	7 (0.6)	19 (3.5)	**157** (5.7)
Holland	22 (5.3)	8 (3.2)	98 (53.0)	11 (6.6)	3 (4.3)	36 (3.2)	28 (5.2)	**206** (7.5)
Spain	6 (1.5)	2 (0.8)	0 (0.0)	54 (32.3)	0 (0.0)	3 (0.3)	3 (0.6)	**68** (2.5)
Switzerland	20 (4.9)	14 (5.5)	8 (4.3)	4 (2.4)	36 (51.4)	15 (1.3)	34 (6.3)	**131** (4.8)
United Kingdom	71 (17.2)	43 (17.0)	40 (21.6)	29 (17.4)	9 (12.9)	977 (86.5)	68 (12.7)	**1237** (44.9)
Germany	34 (8.3)	37 (14.6)	27 (14.6)	25 (15.0)	15 (21.4)	40 (3.5)	320 (59.7)	**498** (18.1)
Total	**412** (15.0)	**253** (9.2)	**185** (6.7)	**167** (6.1)	**70** (2.5)	**1130** (41.0)	**536** (19.5)	**2753** (100)

Source: SDC International M&A Database.

Notes:
The percentage in each column is w.r.t. the total of the column (e.g., 61.4=253/412).
The percentage in the bottom line is w.r.t. the sum of all the totals (e.g., 15.0=412/2753).
The cells in grey represent domestic investments.

4. ANALYSIS OF INVESTMENT FLOWS BY SECTOR

In this section, we focus on a detailed examination of the industries involved in the transactions. The purpose of this exercise is twofold. First, we want to cast a light on the magnitude of flows of investment in each of the main sectors composing the heterogeneous universe of the chemical industry. Second, we assess whether there are differences between the intra-regional and international investment strategies of acquirers in terms of sectors involved. Given the weight of Western Europe and North America, we focus on mergers and acquisitions where either the target or the parent company originates in one of these two regions.

Table 5. Number of M&As by Industry

	Industry	Number	%
1°	Chemicals:		
	Agricultural Chemicals	433	2.2
	Pharmaceuticals	2605	13.3
	Inorganic Chemicals	1131	5.8
	Organic Chemicals	520	2.7
	Paints	425	2.2
	Soap and Detergent	959	4.9
	Synthetic Resins and Rubber	895	4.6
	Miscellaneous Chemicals	897	4.6
	Total Chemicals	**7865**	**40.3**
2°	Oil and Gas Extraction	**6370**	**32.6**
3°	Rubber and Plastic:		
	Plastic	1965	10.1
	Rubber	365	1.9
	Packing and Sealing Devices	142	0.7
	Tyres	151	0.8
	Total Rubber and Plastic	**2623**	**13.5**
4°	Petroleum Refining	**727**	**3.7**
5°	Distribution	**686**	**3.5**
6°	Measuring and Analyzing Instruments	**370**	**1.9**
7°	Engineering and Research	**341**	**1.7**
8°	Industrial and Commercial Machinery	**274**	**1.4**
9°	Electric and Gas Services	**263**	**1.3**
	Total	**19519**	**100**

Source: SDC International M&A Database.

Notes:
Chemicals and Rubber and Plastic are divided into sub-sectors, at 3-digit level.
Industries in Light-Grey cells are named Core Chemicals (CCs).
Industries in Dark-Grey areas are named Vertically Related Industries (VRIs).

Using the 2-digit code of these firms we are to identify the sectors involved. Recall that our database reports all the transactions where the acquiring and/or the target firms belong to one of the following industries: chemicals (SIC 28), petroleum refining (SIC 29), rubber and plastic (SIC 30) and oil and gas extraction (SIC 13). This implies that target firms do not necessary belong to one of these sectors. Indeed, target firms listed in our dataset belong to more than 70 different industries. As many of these industries only record a small number of transactions, Table 5 reports the sectors for which at least 200 acquisitions have been registered. There are 10 different industries satisfying this condition.

Two among of these 10 industries, chemicals and rubber and plastic, are too broad at the 2-digit level. Thus, we have broken them down into their 3-digit equivalents. For the sake of simplicity, we have grouped wholesale trade of durables and that of non-durables into a single sector labelled "Distribution". Accordingly, we ended-up with the following 19 (2 & 3 digit) industries: Agricultural Chemicals, Pharmaceuticals, Inorganic Chemicals, Organic Chemicals, Paints, Soap and Detergents, Synthetic Resins and Rubber, Miscellaneous Chemicals, Oil and Gas Extraction, Plastic, Rubber, Packing and Sealing Devices, Tyres, Petroleum Refining, Distribution, Measuring and Analysing Instruments, Engineering and Research, Industrial and Commercial Machinery, and Electric and Gas Services.

All the sectors listed above fall into two main groups that we label Core Chemicals (CCs) and Vertically Related Industries (VRIs), respectively. The first one groups Chemicals, Plastic and Rubber, Oil and Gas Extraction and Petroleum Refining. The second one is formed by industries that are vertically related to Core Chemicals such as Engineering and Research, Industrial and Commercial Machinery, Measuring and Analysing Instruments (upstream Industries), and Electric and Gas Services and Distribution (downstream Industries).

Before analysing the distribution of intra-regional and international M&As for each of the 19 sectors listed above, we first present some data on the total number of active firms in Europe and North America in these industries. This information is relevant given that the absolute number of M&As has to be related to the total number of potential targets to be meaningful. We use the number of establishments for the year 1992, as reported in the OECD Industrial Structure Statistics database. The number of establishments can be considered a reliable proxy for the number of firms given that the average number of establishments per firm is likely to be rather similar in the two areas. Industries are grouped following the ISIC classification, which is similar but not equal to the SIC classification we are using in our analysis for CCs. As far as VRIs are concerned, we have reported data only for those industries that are related to the chemical sector. The figures reported in Table 6 indicate that population size across the two regions does not vary much. As far as Core Chemicals are concerned, there is a slight predominance of North America in Fertilisers and Pesticides (that is, Agricultural Chemicals) and Chemical Products n.e.c., while the reverse is true for Synthetic Resins and Paints. The comparison is less clear-cut for VRIs, probably because of the lack of suitable data. However, the overall picture is that the population of European and American establishments does not differ much. This means that the absolute number of intra-regional and international M&As can be directly compared without adjusting the figures for the relative number of total targets in the two areas.

Table 6. Number of Establishments in Western Europe and North America

	Total	Western Europe		North America	
Industry (ISIC)		Number	%	Number	%
Core Chemicals					
Fertilisers & Pesticides	2113	898	42.5	1215	57.5
Pharmaceuticals	4138	2221	53.7	1917	46.3
Basic Industrial Chemicals	5336	2783	52.2	2553	47.8
Paints	4383	2493	56.9	1890	43.1
Synthetic Resins	1889	1146	60.7	743	39.3
Soap, Detergent & Perfumes	4524	2106	46.6	2418	53.4
Chemical Products, nec	9501	3852	40.5	5649	59.5
Plastic	35602	17072	48.0	18530	52.0
Rubber	6213	3175	51.1	3038	48.9
Petroleum Refining	3539	1037	29.3	2502	70.7
Vertically Related Industries					
Indus. & Commer. Machinery					
Engines & turbines	2155	1683	78.1	472	21.9
Metal and wood working mach.	27302	11241	41.2	16061	58.8
Computing & office equipment	5737	2349	40.9	3388	59.1
Special industrial machinery	21841	11988	54.9	9853	45.1
Measur. And Analys. Instrument					
Photographic and optical goods	4517	1530	33.9	2987	66.1
Professional equipment	18082	5664	31.3	12418	68.7

Source: OECD Industrial Structure Statistics (1995).

Note:

Figures refers to year 1992. Data about number of establishments in Fertilisers and Pesticides and Chemical Products n.e.c., were not available for Germany. No data about number of establishments were available for France. We have obtained the missing values for Germany and France using the data about number of employees. Comparing this latter variable with other European countries, we have inferred reasonable figures for the missing values.

We can now compare the flow of intra-regional and cross-border investments between European and Northern American companies, for the 19 industries listed in Table 5. Recall that our analysis of investment flows by industry is based on the SIC code of the target firm. Given that intra-regional acquisitions in "Oil and Gas Extraction" companies in North American are over-represented compared to Europe (about 50% of domestic transactions in US compared to 15% for Europe), Table 7 shows the distribution of M&As without this sector.

As far as intra-regional M&As are concerned (columns 1 and 2 of Table 7), we observe some slight differences in the relative importance of core industries. In North America more than 25% of the acquisitions have involved pharmaceutical companies compared to 18.7% for Europe. This is another reflection of the attractiveness of US assets in the pharmaceutical and biotech sectors. By contrast, Plastic and Rubber, and Inorganic Chemicals account for a higher percentage of intra-regional M&As in Europe compared to the US. The figures in columns 3 and 4 indicate that most of the differences identified above vanish for international M&As. Indeed, there are striking similarities in the distribution of cross border M&As between Europe and North America. The relative weight of most of the industries differs by less than 1 percentage point. This finding suggests that, in most sectors, European and US firms are similarly endowed, since international M&As do not seem to reflect differences that could be explained by traditional notions of comparative advantage. By contrast, the degree of symmetry suggests that oligopolistic rivalry may be an important force driving M&As across the Atlantic. We return to this point below.

In the bottom part of the table we analyse the relative importance played by up-stream and down-stream industries. While the number of domestic M&As seems to be pretty much the same for the North American region, "Distribution" is prominent in Europe where more than 50% of the acquisitions have been targeted at firms of this sector. Once again, when we look at the figures for international M&As, it is hard to discern differences in the relative magnitude of the industries listed. This is further evidence that US and European chemical firms are similarly endowed.

If we compare the figures of Europe and North America, we find that the absolute number of intra-regional acquisitions in North America is constantly higher in most of the cases (Inorganic Chemicals being the only exception). The four industries where the relative gap is noticeable (apart from Oil and Gas Extraction) are Pharmaceuticals, Electric and Gas Services, Engineering & Research, and Measuring & Analysing Instruments. Three of these sectors are high-tech, confirming the attractiveness of US assets. It also suggests that US firms are quite active when it comes to M&As in these sectors.

By contrast, figures on international operations show that the absolute number of cross-border M&As is constantly higher for Europe, with the exclusion of Paints and Industrial and Commercial Machinery. Quite interestingly, these latter two are sectors of relative European strength.

Finally, the last two columns of Table 7 present the ratio between intra-regional and international M&As by sector. Bearing in mind that there is probably a US bias in the data, there are some interesting features emerging from these figures. First, the relative weight of domestic M&As is higher in North America than in Europe. For CCs, the average ratio in North America is 10.2 against 3.7 in Europe, and 7 against 2.3 for VRIs. This is further evidence that the European chemical industry is more outward oriented. It could also reflect the fact that the "market for firms" is less developed in Europe. As far as CCs are concerned, Pharmaceuticals is the sector with the highest weight of foreign investments for Europe and one of the greatest for North America. It is also worth noting that, for Pharmaceuticals, acquisitions of US assets by European firms is large compared to intra-European transactions (column 5 of Table 7). For transactions in CCs, the purchase of US Pharmaceutical assets by European firms is large, both in absolute terms (200 transactions out of a total of 708) as well as in relative terms (lowest ratio of intra-regional to international transactions). This suggests that in Pharmaceuticals, European firms rely on the acquisition of US assets to maintain their knowledge base. By contrast, Paints is the one with the lowest relative weight of trans-Atlantic transactions. This evidence corroborates the findings reported in the case studies of Brusoni (1998) and Nightingale (1998).

It is interesting to note that the number of international acquisitions for "Engineering & Research" and "Measuring & Analysing Instruments" by European parents does not differ much from intra-regional transactions. This may be a reflection of the comparative strength of some European firms in these two sectors.

The general picture that we get from this second set of results is that there are no marked differences in the investment process undertaken by European and Northern American acquirers. Although the absolute number of domestic investment appears to be higher in North America and the absolute number of foreign investment is larger in Europe, the relative importance played by the different industries composing the universe of chemicals is pretty much the same in these two regions. The striking feature is the degree of similarity characterising M&As across the Atlantic. The sectoral distribution of international M&As is virtually symmetric. Two non-exclusive explanations could account for this finding. First, that the oligopolistic structure of the industry results in bilateral M&A flows. This is reminiscent of the "reciprocal dumping models" of the so-called "New International Trade" (NIT) theory where Cournot oligopolists invade each other markets. Second, the symmetry of flows may reflect the existence of similar firm specific advantages of the

European and North-American firms. This would reinforce the oligopolistic interactions mentioned above. The results presented in this section also give some support to the idea that European companies acquire firms in the US in order to source leading edge technology in Pharmaceuticals and Biotech.

Table 7. Distribution of Intra-Regional and International M&As by Industry (without Oil and Gas Extraction)

	Intra-Regional				International				Ratios	
	Western Europe (1)		North America (2)		W Europe to N.America (3)		N. America to W.Europe (4)		Intra-Reg./Intern. (1)/(3)	(2)/(4)
Industry	Number	%	Number	%	Number	%	Number	%	W. E	N. A.
Core Chemicals										
Agricultural Chemicals	120	3,5	193	3,9	28	3,1	17	2,4	4,3	11,4
Pharmaceuticals	648	18,7	1280	26,1	247	27,2	200	28,2	2,6	6,4
Inorganic Chemicals	408	11,8	390	7,9	126	13,9	94	13,3	3,2	4,1
Organic Chemicals	142	4,1	234	4,8	46	5,1	36	5,1	3,1	6,5
Paints	169	4,9	178	3,6	22	2,4	25	3,5	7,7	7,1
Soap, Detergent, Perfumes	305	8,8	452	9,2	55	6,1	55	7,8	5,5	8,2
Synth. Resins & Rubber	310	8,9	321	6,5	84	9,2	71	10	3,7	4,5
Miscellaneous Chemicals	283	8,2	356	7,3	92	10,1	64	9	3,1	5,6
Tyres	47	1,4	61	1,2	13	1,4	7	1	3,6	8,7
Packing and Sealing	43	1,2	64	1,3	14	1,5	11	1,6	3,1	5,8
Plastic	666	19,2	888	18,1	120	13,2	83	11,7	5,6	10,7
Rubber	141	4,1	150	3,1	27	3	19	2,7	5,2	7,9
Oil and Gas extraction										
Petroleum Refining	190	5,5	339	6,9	35	3,9	26	3,7	5,4	13
Total Core	3472	100	4906	100	909	100	708	100	3,7	10,2
Vertically Related Indus.										
Electric and Gas Services	47	9,5	185	18,6	14	6,5	11	7,7	3,4	16,8
Engineering and Research	64	13	188	18,9	52	24	24	16,9	1,2	7,8
Indus. & Commer. Mach.	65	13,2	148	14,9	22	10,1	26	18,3	3	5,7
Meas. & Analys. Instrum.	64	13	191	19,2	62	28,6	36	25,4	1	5,3
Distribution	253	51,3	284	28,5	67	30,9	45	31,7	3,8	6,3
Total Vertical	493	100	996	100	217	100	142	100	2,3	7

Source: SDC International M&A Database.
Note:
The percentage is w.r.t the total of either core or vertically related (e.g., 3.5=120/3475).

5. DIVERSIFICATION STRATEGIES

We now turn to the diversification strategies and the extent of ownership control over acquired firms. In particular, we compare intra-regional and international M&As, and attempt to provide explanations for these differences when they exist. We pay particular attention to the R&D intensity of the industries, and assess whether this results in different diversification strategies.

Using the 4-digit SIC code of target and acquiring companies and the distinction between Core Chemicals (CCs) and Vertical Related Industries (VRIs), we identify six different diversification strategies, as depicted in Figure 1.

Figure 1. Diversification Strategies

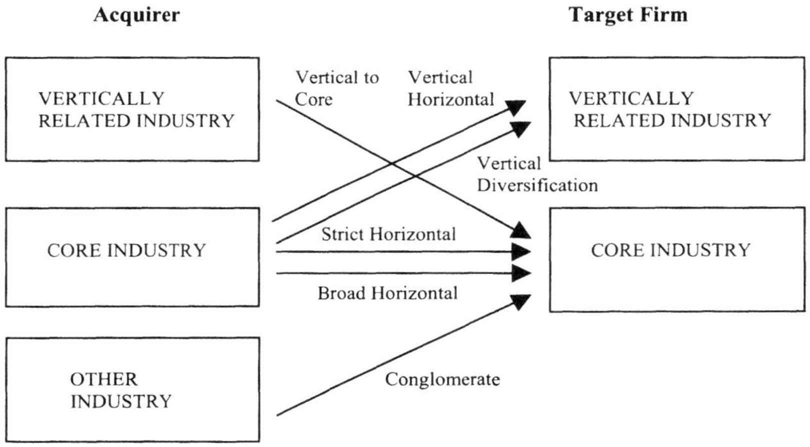

When both firms belong to CCs, we identify two types of acquisitions labelled Strict and Broad Horizontal. Strict Horizontal: whenever the ultimate acquirer and the target company have the same 4-digit SIC, and Broad otherwise. Second, when the acquirer has a product line in CCs and the target firm belongs to one of the VRIs, we have a case of vertical integration. We also consider two distinct sub-groups. Vertical-Horizontal when the acquiring company was already present in the target firm's industry, and Vertical-Diversification otherwise. Third, when the target firm belongs to CCs and the parent belongs to one of the five VRIs, we label this transaction as Vertical-to-Core. Finally, when the acquirer comes from an unrelated industry, we have a case of Conglomerate diversification.

Table 8 shows the relative weight of these diversification strategies for intra-regional and international transactions. The role played by conglomerate M&As is much higher in domestic transactions compared to international M&As. This reflects the fact that conglomerate acquisitions are, in many instances, aimed at diversifying risk and smooth revenue flows over the business cycle. In this context, domestic investments are attractive, since acquirers are likely to possess better knowledge of local markets.

The column labelled Total Horizontal indicates that the relative importance of this strategy is larger for international transactions compared to domestic ones. A related finding is that horizontal acquisitions play a major role for North American acquirer, both in intra-regional (62.4% vs. 56.6%) and cross-border (69.9% vs 66.8%) M&As. Also, most of the transactions in Core Chemicals involve firms operating in the same product line, independently of the region considered. Indeed, the number of Strict Horizontal transactions is much higher than that of Broad Horizontal.

As far as vertical integration is concerned, the relative importance of this diversification strategy is higher for international M&As compared to domestic ones. The number of Vertical-Diversification transactions is consistently higher than Vertical-Horizontal M&As. Moreover, Table 8 shows that the number of Vertical-to-Core transactions is pretty much the same as those labelled Vertical Diversification. This means that vertical integration is the result of two-way flows: chemical firms expanding their activity up-stream and down-stream, but also non-chemical firms expanding in CCs.

Finally, it is interesting to note that the ratio between horizontal and vertical acquisitions does not differ that much between domestic and international transactions. For example, in the case of Europe the ratio is 2,8 (56,6/20,4) for intra-regional M&As, and 2,7 (66,7/24) for cross-border M&As. Similar evidence is obtained for North American companies. This is further evidence of the industry's globalisation: overall, international strategies do not appear to differ much from domestic ones.

Comparing the relative weight of Strict and Broad Horizontal diversification strategies for all the nineteen industries we are considering, it emerge that, for North-American firms, Strict Horizontal M&As are more important in Pharmaceuticals and Paints, both for domestic and international investments. R&D investments in Pharmaceuticals are large and may require many years to mature, thus making this strategy attractive to acquire core competencies in that sector. As far as Paints are concerned, Europe enjoys comparative advantage in this sector (see Brusoni (1998)). Instead of intending to produce skills themselves, North-American firms appear to be more inclined to turn to M&As to broaden their capabilities to this field. For European firms, Strict Horizontal transactions have a greater importance in Tyres and Pharmaceuticals industries, both for domestic and foreign M&As.

Table 8. Diversification Strategies across the Atlantic

	Target's Region: North America								
Acquirer Region:	Total	Conglo-merate	Horizontal Total	Broad	Strict	Vertical Total	to Core	Horiz.	Diver.
N. Amer.	10485	2182	6547	1311	5236	1756	762	189	805
	(100)	(20.8)	(62.4)	(12.5)	(49.9)	(16.7)	(7.3)	(1.8)	(7.7)
W. Europe	1340	124	895	357	538	321	104	24	193
	(100)	(9.3)	(66.8)	(26.6)	(40.1)	(24.0)	(7.8)	(1.8)	(14.4)
Total	11825	2306	7442	1668	5774	2077	866	213	998
	(100)	(19.5)	(62.9)	(14.1)	(48.8)	(17.6)	(7.3)	(1.8)	(8.4)
	Target's Region: Western Europe								
N. Amer.	1075	94	750	242	508	231	89	30	112
	(100)	(8.7)	(69.8)	(22.5)	(47.3)	(21.5)	(8.3)	(2.8)	(10.4)
W. Europe	4614	1060	2611	941	1670	943	450	56	437
	(100)	(23.0)	(56.6)	(20.4)	(36.2)	(20.4)	(9.8)	(1.2)	(9.5)
Total	5689	1154	3361	1183	2178	1174	539	86	549
	(100)	(20.3)	(59.1)	(20.8)	(38.3)	(20.6)	(9.5)	(1.5)	(9.7)

Source: SDC International M&A Database.
Note:
The percentage below the number is w.r.t the total of the line (e.g., 20.8=2182/10485).

Table 9 presents a simple analysis of the percentage of ownership acquired by American and European firms. The extent of majority M&As is pervasive. In most cases, Majority accounts for more than 80% of transactions, irrespective of the region concerned or of the nature the diversification strategy pursued. The only exception is Conglomerate where the percentage of Minority ownership is quite high. For example, almost 50% of American domestic investments for this diversification strategy fall in this group. This confirms that many conglomerate M&As are undertaken to smooth revenue flows, which do not necessarily require full managerial control of the subsidiary.

Recall that transactions in CCs that do not share any product line are represented in Broad Horizontal. The same idea applies to Vertical-Diversification and Vertical-to-Core. We would expect Minority acquisitions to be more common for Broad Horizontal, Vertical-Diversification and Vertical-to-Core because of the acquirer's supposed lack of experience in these markets.

The evidence provided in Table 9 does not support this hypothesis for Horizontal diversification strategies. Minority is constantly lower for Broad than Strict Horizontal, whatever the market entered. The results for Vertical integration are mixed. Minority plays a major role both in absolute and relative terms in Vertical-Diversification and Vertical-to-Core for investments originating in North America (first two columns). However, this is not the case for M&As undertaken by European acquirers.

In the same line, Minority may have been expected to be more common for international transactions compared to domestic because of the acquirer's lesser familiarity with foreign markets. This is however not the case, since Minority is almost always higher for domestic than cross-border M&As. A possible explanation for this latter finding is that multinational firms prefer to retain full managerial control of the their foreign subsidiaries. This is consistent with the idea that multinationals seek to avoid the diffusion of their firm specific assets to actual or potential competitors when going abroad.

Table 9. Percentage of Ownership by Type of Diversification Strategy

Diversification Strategy	**% of Owner-Ship**	**North America to** North America Numb.	%	**North America to** Western Europe Numb.	%	**Western Europe to** North America Numb.	%	**Western Europe to** Western Europe Numb.	%
Strict	majority	2503	81.7	302	85.8	328	88.9	1012	87.9
Horizontal	minority	561	18.3	50	14.2	41	11.1	139	12.1
	Total	**3064**	**100**	**352**	**100**	**369**	**100**	**1151**	**100**
Broad	majority	881	92.0	169	93.9	260	92.9	597	89.4
Horizontal	minority	77	8.0	11	6.1	20	7.1	71	10.6
	total	**958**	**100**	**180**	**100**	**280**	**100**	**668**	**100**
Conglomer.	majority	874	55.5	42	65.6	49	60.5	536	72.2
	minority	702	44.5	22	34.4	32	39.5	206	27.8
	total	**1576**	**100**	**64**	**100**	**81**	**100**	**742**	**100**
Vertical –	majority	139	94.6	20	95.2	16	80.0	36	92.3
Horizontal	minority	8	5.4	1	4.8	4	20.0	3	7.7
	total	**147**	**100**	**21**	**100**	**20**	**100**	**39**	**100**
Vertical –	majority	501	87.4	75	92.6	133	86.9	264	88.0
Diversific.	minority	72	12.6	6	7.4	20	13.1	36	12.0
	total	**573**	**100**	**81**	**100**	**153**	**100**	**300**	**100**
Vertical	majority	467	88.8	69	92.0	75	88.2	282	85.7
to Core	minority	59	11.2	6	8.0	10	11.8	47	14.3
	total	**526**	**100**	**75**	**100**	**85**	**100**	**329**	**100**

Source: SDC International M&A Database.

The final step of our analysis consists in assessing whether the degree of ownership control differs depending on the R&D intensity of the industry involved. This classification is based on our analysis in chapter 1 of this book. The results are presented in Table 10.

First, there is no industry for which Minority acquisition plays a primary role. Moreover the relative weight of Minority is sensibly higher for domestic M&As, with the only exception involving European firms in Pharmaceuticals, Measuring and Analysing Instruments, and Engineering and Research. That is, European firms have a higher propensity to acquire minority stakes in US assets in sectors that are R&D intensive. This suggests that technology sourcing may be occurring. By not wrestling full control of the operation, the parent company supposedly wishes to allow existing management, and more generally, human resources a certain degree of autonomy. Coupled with the fact that multinationals tend to prefer majority M&As, this suggests that minority stakes in R&D intensive sectors are motivated by the desire to acquire core competencies that can not be bought of the shelf, or built from scratch at a reasonable cost. This is also consistent with evidence that, in the case of Pharmaceuticals, the research base has been shifting towards the US.

The overall picture is one of similarity between European and US firms' M&A strategies. The exception to this general finding pertain to high-tech sectors. In the latter case, European firms' strategies is consistent with the hypothesis of "technology sourcing", i.e. part of the acquisitions of US assets undertaken by European firms is driven by the necessity to maintain and expand their knowledge base.

Table 10. Percentage of Ownership by Industry

	North American to				Western Europe to			
	N. America		W. Europe		N. America		W. Europe	
Industries	Major	Minor	Major	Minor	Major	Minor	Major	Minor
High R&D:								
Agricultural Chemicals	117	16	6	0	24	0	76	5
	(88.0)	(12.0)	(100)	(0.0)	(100)	(0.0)	(93.8)	(6.2)
Pharmaceuticals	541	275	35	0	116	42	408	71
	(66.3)	(33.7)	(100)	(0.0)	(73.4)	(26.6)	(85.2)	(14.8)
Organic Chemicals	174	57	13	3	34	1	74	22
	(75.3)	(24.7)	(81.3)	(18.8)	(97.1)	(2.9)	(77.1)	(22.9)
Synth. Resins & Rubber	207	32	41	3	66	1	189	20
	(86.6)	(13.4)	(93.2)	(6.8)	(98.5)	(1.5)	(90.4)	(9.6)
Indus. & Commer. Mach.	105	9	17	1	19	2	44	7
	(84.7)	(7.3)	(94.4)	(5.6)	(90.5)	(9.5)	(86.3)	(13.7)
Measur. & Analys. Instr.	128	19	22	2	44	7	39	4
	(87.1)	(12.9)	(91.7)	(8.3)	(86.3)	(13.7)	(90.7)	(9.3)
Engineering and Research	108	20	13	3	18	12	36	4
	(84.4)	(15.6)	(81.3)	(18.8)	(60.0)	(40.0)	(90.0)	(10.0)
Medium R&D:								
Paints	91	39	22	0	17	1	108	16
	(70.0)	(30.0)	(100)	(0.0)	(94.4)	(5.6)	(87.1)	(12.9)
Plastic	561	87	66	6	92	9	478	21
	(86.6)	(13.4)	(91.7)	(8.3)	(91.1)	(8.9)	(95.8)	(4.2)
Rubber	81	16	12	3	21	0	91	10
	(83.5)	(16.5)	(80.0)	(20.0)	(100)	(0.0)	(90.1)	(9.9)
Oil and Gas Extraction	2187	652	90	48	91	38	244	162
	(77.0)	(23.0)	(65.2)	(34.8)	(70.5)	(29.5)	(60.1)	(39.9)
Petroleum Refining	136	65	21	0	30	0	84	31
	(67.7)	(32.3)	(100)	(0.0)	(100)	(0.0)	(73.0)	(27.0)
Low R&D:								
Soap, Deterg. & Perfumes	236	66	41	3	41	2	202	25
	(78.1)	(21.9)	(93.2)	(6.8)	(95.3)	(4.7)	(89.0)	(11.0)
Inorganic Chemicals	206	45	65	6	90	4	243	48
	(82.1)	(17.9)	(91.5)	(8.5)	(95.7)	(4.3)	(83.5)	(16.5)
Distribution	189	11	35	0	59	1	166	13
	(94.5)	(5.5)	(100)	(0.0)	(98.3)	(1.7)	(92.7)	(7.3)

Source: SDC International M&A Database.

Note:
The percentage below the number (e.g., 88.0=117/(117+16)).

6. CONCLUSIONS

This chapter has examined some general patterns of Mergers and Acquisitions in the chemical industry between 1985 and 1997. We have focused our attention on the investment strategies of American and European firms, both domestic and international. We provide stylised facts on investment flows, and attempt to identify differences across target areas and industries. Besides, we provide an analysis of the mode of entry and the extent of ownership control.

As emphasised by Arora et al.(1998), there has been an increasing globalisation process for this industry and Mergers and Acquisitions are the main tool used by parent companies to expand. Bearing in mind that there may be a bias in the data, the picture that emerges is that Europe and America are the main source and destination of investments through mergers and acquisitions, while Japanese companies appear to play a more modest role. Another interesting pattern that emerges is that, apart from high-tech sectors, there are no remarkable differences between domestic investments and foreign direct investments undertaken by industrialised countries in terms of the industries involved. In fact, the sectoral distribution and number of M&As undertaken by US and European firms is almost perfectly symmetric. This suggests that oligopolistic rivalry is an important driving force behind M&As in the chemical industry.

As far as the mode of entry is concerned, a marked pattern is the proliferation of vertical acquisitions. One possible explanation of this evidence is that the increased competition and globalisation of the market push most of the firms to diversify their activities up-stream and down-stream, in particular in distribution and in research activities. In addition, we show that there is a prevalence of wholly-owned acquisitions. Last, we report suggestive evidence that indicates that European firms have a higher propensity to acquire minority stakes in sectors that are high R&D intensive. This suggests that some technology sourcing may be occurring. Moreover, it appears that European pharmaceutical firms are increasingly relying on acquisition of US assets to maintain their knowledge base.

7. REFERENCES

Arora, A., Gambardella, A. and Garcia-Fontes, W. (1998), "Investment Flows of Large Chemical Firms", paper presented at the Conference on Chemical Industry TSER "From Science to Products held in Stuttgard on the 11th – 12th December 1998 (http://puna.upf.es/tser.html).

Brusoni S. (1998), "Watching Paint Dry? Environmental Regulation and Innovative Dynamics in the Paint Industry", In Gambardella A. and Garcia-Fontes (eds.) 'The European Chemical Industry: Innovation, Performance and Competitiveness', chapter7.

Dunning, J., (1981), "International Production and the Multinational Enterprise", George Allen and Unwin, London.

Fosfuri, A., and Motta, M., (1999), "Multinationals without Advantages", Scandinavian Journal of Economics, Vol. 101, pp. 617-630.

Kang, N., and Johansson, S., (2000), "Cross-border Mergers and Acquisitions: Their Role in Industrial Globalisation" STI Working Papers 2000/1, OECD Paris.

Kogut, B., and Chang S. (1991). "Technological Capabilities and Japanese Foreign Direct Investment in US", Review of Economics and Statistics, pp. 401-413.

Marín, P., and Siotis, G., (2002), "Innovation and Market Structure: An Empirical Evaluation of the "Bounds" Approach in the Chemical Industry", CEPR DP #3162, January.

Markusen, J., (1995), "The Boundaries of Multinational Enterprises and the Theory of International Trade", Journal of Economic Perspectives, Vol. 9, pp. 169-189.

Neven, D., and Siotis, G., (1996), "Technology Sourcing and FDI in the EC: an Empirical Evaluation", International Journal of Industrial Organization, Vol. 14, pp. 543-560.

Nightingale, P., (1998), "The Competitiveness of the European Pharmaceutical Industry" paper presented at the Conference on Chemical Industry TSER "From Science to Products" held in Stuttgart on the 11th – 12th December 1998,

Serapio, M.G. And Dalton, D.H. (1999), "Interindustry technological flows and productivity growth", Review of Economics and Statistics LXIV, pp 627-634.

Chapter 3

CORPORATE RESTRUCTURING AND R&D: A PANEL DATA ANALYSIS FOR THE CHEMICAL INDUSTRY

Ashish Arora, Marco Ceccagnoli and Marco Da Rin
Carnegie Mellon University, INSEAD and Turin University, ECGI and IGIER

Abstract: We provide a novel approach to the existing literature on the effects of restructuring on R&D investment by focusing on a single industry, chemicals. The chemical industry is very research intensive and has experienced thorough restructuring since the early 1980s. By focusing on a single industry we are able to identify the technological and R&D features of its segments. This is important, since there is evidence that restructuring affects R&D differently in businesses with different technological features. However, no study so far has provided a systematic inquiry into this link. Using a panel of 535 European, American, and Japanese firms for the years 1987-1997 we find restructuring to be an important component in the observed changes in R&D intensity. We show that restructuring affects R&D both through changes in size and through changes in the composition of business portfolios, and that these effects differ across industry segments.

Key words: Restructuring Process; Innovation; Chemical Industry

F. Cesaroni et al. (eds.), R&D, Innovation and Competitiveness in the European Chemical Industry, 69-90.

1. INTRODUCTION

Investment in research and development (R&D) is an important source of long-term growth for industrial economies, and the economics of R&D is an intensely researched topic. Much attention has gone into the determinants of the R&D investment decision, both theoretically and empirically. In mature economies, firms and industries increasingly evolve through restructuring (Mitchell and Mulherin (1996)). Restructuring at the firm level entails changes in the composition of both capital and labor, and in particular the divestiture and acquisition of productive assets. Restructuring at the level of the industry entails the entry and exit of firms through takeovers, mergers and acquisitions, i.e. sales and purchases of whole businesses. It is then important to understand what effect restructuring has on R&D. However, this topic has received relatively little attention so far. In this chapter we provide a study of firm-level restructuring which sheds new light on how firms change their R&D investment as a result of changes in their business portfolios.

In the late 1980s, a number of empirical studies have analyzed the effects of restructuring on R&D. These efforts were motivated by the suspicion that the increased takeover activity among U.S. firms in the early 1980s might have induced 'managerial myopia' (Stein (1988)), and thus a decrease in long-term investments such as R&D. The brunt of this exercise was done by Bronwyn Hall (1988, 1990). Using information on U.S. listed manufacturing companies, she examined the determinants of acquisitions, making some interesting points. First, acquisitions by public companies do not result in any systematically significant decline in R&D spending, nor there is any significant difference in R&D spending between firms actively restructuring through acquisitions and inactive firms. Moreover, the difference in R&D intensity—defined as the ratio of R&D to sales—between acquiring and acquired firms is statistically insignificant, while the population of acquiring firms experienced a decline in R&D intensity (relative to their industry) over the 1980s. Second, R&D intensity marginally declines following an acquisition by a private company (a 'going private' transaction, such as a leveraged or management buy-out), a type of acquisition which occur prevalently in industries where R&D intensity is low to start with. Third, whether induced by the acquisition or by other events, large increases in leverage (i.e. larger than 50%) results in significant decreases in R&D intensity. For firms not engaged in going private transactions this effect was sizeable: on average about 0.8 percent of sales. Finally, R&D is a poor predictor of acquisitions, with the notable exception of the cases when a firm was being taken private.

Hall (1994) makes a further point, which motivates our study. She shows that the pressure to restructure is not constant across industries, but depends on the pace of technological change, a fact also documented by Blair and Litan (1990) and Blair and Schary (1993). Hall found that the four categories used by Chandler (1990) to partition manufacturing industries—high-, low- and stable- (with short and long lifetime horizons) technology—are good predictors of acquisition probabilities. Firms with low- or stable-, short horizon, technology are the likeliest candidates for an acquisition, especially through a "going private" transaction. Firms with stable, long horizon, technology, among which are chemicals (excluding pharmaceuticals), are likeliest to take on large leverage and to reduce physical investment.

These contributions still represent the locus classicus for the analysis of how corporate restructuring affects R&D investment. However, they do not address some important questions. In particular, they do not look into the effect of the reallocation of asset portfolios across different segments within an industry. The intensity of restructuring varies across industries, as does its effect on R&D investment. Since most industries are typically made up of technologically heterogeneous segments, variation in the effects of restructuring on R&D is likely to be experienced also within an industry. It is then important to bring the analysis at the level of the industry and to look into how R&D reacts to restructuring at both segment and firm level. In particular, firm level data can be exploited to relate corporate restructuring to changes in R&D investment taking into account firms' technological (re)positioning. This is our contribution. Simply put, we can observe acquisition or divestiture from different segments, and measure the impact it has on R&D, and further, how that impact varies by industry segment.

We focus on a single industry, chemicals, and study the impact of restructuring on firm R&D. There are several important reasons to focus on chemicals:

(1) First, the chemical industry is R&D intensive, so that changes in a firm's industrial portfolio may require significant changes in its R&D strategy;

(2) Second, there have been some important and clearly identifiable shocks to the industry since the 1970s. The oil shock exposed overcapacity in petrochemicals; the rise of biotechnology severed the link between chemicals and pharmaceuticals and created life sciences as a separate industry; the dissemination of maturing process technology worldwide made 'specialty' chemicals more lucrative than 'commodity' chemicals. The far-reaching nature of these shocks has resulted in continued pressure for restructuring, giving us a precious chance to study the effects of ongoing structural changes;

(3) Third, focusing on chemicals allows us to identify several segments with distinct technological characteristics, and thus to assess precisely the nature of restructuring;

(4) Lastly, chemicals have been a truly global industry since long; hence, looking at chemicals allows us to grasp effects which go beyond national idiosyncrasies.

We thus provide a fresh and fine-grained perspective. Unlike most of the previous work in this area, we analyze not only acquisitions, but also divestitures. Also, we consider not just acquisitions of entire firms, but also of single divisions and businesses. Since the bulk of industrial restructuring in chemicals—as well as many other industries with multi-business firms—involves assets sales and divestitures at the business or product level, our analysis provides further insights than the earlier literature on acquisitions. Moreover, we consider a sample of firms from the U.S., Europe, and Japan, so as to get a broader view of the dynamics of corporate restructuring, a view consistent with the global nature of the chemical industry. And we explore the effects of restructuring for the 1990s, which was an extremely intense period of restructuring in chemicals.

We assemble a unique database which covers the years 1987 to 1997, and which contains company-level information on 535 North American, European and Japanese chemical firms. While our sample is biased towards North American firms, which constitute 58% of the sample, it nonetheless includes most of the large chemical companies from Europe and Japan.

Our main result is that restructuring does matter for R&D investment. Net acquisitions in R&D intensive industry segments have a positive and significant effect on R&D investment, a result robust to different specifications and samples. Notice that, by looking into a single industry and at its segments, we are able to get a finer appraisal of the effects of changes in portfolio composition on R&D. For example, we find that the significance of net acquisitions varies across samples. We also find that the elasticity of R&D with respect to sales is less than one, and varies widely across industry segments. Financial variables like debt or cash flow do affect R&D, but not the effect that restructuring has on R&D. This result complements Hall's (1990) finding that the effect of leverage on R&D intensity does not change whether a firm is an active acquirer or not.

2. DATA DESCRIPTION

Our sample is based on the publicly traded North American, European and Japanese manufacturing firms included in Compustat's Global Vantage, a database which collects income and balance sheet information (including annual R&D expenditures) on thousands of firms worldwide. We select those firms with primary SIC (Standard Industrial Classification) code in chemical and allied products (SIC 28), energy related (SIC 13, 29, 46), tires (SIC 3011), and other large and diversified companies with significant chemical related activities but primary SIC in different industries, based on Aftalion (1989). We include only firms with at least one year of R&D investment information for the years from 1987 to 1997. The time frame chosen reflects a recent wave of intense restructuring in the industry as a whole, and the fact that reliable data for the period before 1987 are not available on a systematic basis. The resulting sample consists of 535 firms, of which about 58% are North American companies, 21% are Japanese, 13% are European, and 8% are from the UK. Given that we only have R&D data for a subset of the firm years, ours is an unbalanced panel.

We then identify the restructuring deals in which the selected firms where involved by linking Compustat's Global Vantage with restructuring data coming from the Security Data Company's (SDC) World-wide Mergers & Acquisitions database, which covers almost 100,000 deals worldwide since 1985. SDC information for the U.S. is based on SEC filings (8Ks, 10Ks, 10Qs), proxy statements, tender offers, annual reports. For non-U.S. deals SDC relies on corporate reports and on over 200 news sources. We selected about 16,000 world-wide chemical related transactions, by including deals announced between 1987 and 1997 where the primary SIC of either the acquirer or the target company was in chemical and allied products (SIC 28), tires (SIC 3011), or oil, gas, and petroleum refining (SIC 13,29,46). The resulting 535 sample firms are responsible for about a third of the selected worldwide chemical related deals.

Each firm is assigned to one primary segment: Energy, Commodity Chemicals, Life Sciences, Other Chemicals, and Others. The Energy segment (SIC codes 13, 29, 46) is characterized by a low R&D intensity, about 1%, and the major research efforts consists of developing and improving process technologies. Many oil firms are now major chemical producers, particularly of basic petrochemical products such as ethylene and propylene, as well as a variety of industrial solvents. In the 1970s and 1980s, many of these firms further expanded into the production of a variety of plastics such as polyethylene and polypropylene.

Commodity Chemicals (SIC codes 281, 282, 286) refers to chemicals produced in large quantities, and includes major inorganic chemicals such as fertilizers and soda, as well as organic chemicals such as solvents (e.g. pthalic anhydride), basic plastics (e.g. polyethylene) and fibers (e.g. polyester). Products in this segment are typically standard, their demand is highly price sensitive and cyclical, and the technology required for production, though sophisticated, is diffused widely. The average R&D intensity in this segment is low—around 3% in the sample period—although a few firms continue to make major investments in developing new processes and catalyst systems.

The Other Chemicals segment (SIC codes 289 (specialty), 287 (agricultural chemicals), 284 (Soap, Cleaners, and Toilet Goods), and 285 (Paints and Allied Products)) contains a variety of businesses and products, many of which tend to be differentiated and branded, such as adhesive, engineering plastics, food additives, and fine chemicals. Demand is less price elastic than in Commodities or Energy, products are often customized to specific uses, and service and branding are an important source of competitive advantage for firms. Accordingly, research in this segment tends to focus on improving the properties of materials, and its R&D intensity is about 3%.

Life Science (SIC code 283) is the most research intensive segment, consisting mainly of pharmaceutical products, but also agricultural products such as seeds. The advent of recombinant DNA technology and molecular biology has opened up a number of opportunities for product and process innovation. This segment also has a substantial number of small but very research intensive firms, in contrast to the other segments where the smaller firms tend not to invest in research. Over the sample period, the average R&D intensity rises from 9% at the beginning of the period to about 11%, showing a positive trend over time, differently from the other segments, whose R&D intensity is quite stable over time.

Finally, we assign each company to one of four regions: North America (i.e. United States and Canada), Japan, United Kingdom, and Continental Europe . Tables 1 to 3 provide descriptive statistics for our sample. They show that nearly half of the observations come from deals involving North American companies. Japan is relatively under-represented with respect to the size of its economy, while the share of transactions accruing to the UK is larger than that of Europe.

Table 1 : Restructuring volumes and value, by region

Region	Number	Average value ($m)	Observations with value (%)	Number	Average value ($m)	Observations with value (%)
North America	1,635	8.47	43%	776	9.45	55%
Japan	155	1.48	46%	13	0.21	39%
EU*	903	26.99	32%	178	16.94	32%
UK	609	17.07	62%	241	12.89	53%

*: Continental Europe.

Table 2 : Restructuring volumes, by segment and region

		North America	%	Japan	%	EU*	%	UK	%	TOT	%
ENERGY	Acq.	326	58%	5	0.9%	148	26%	82	15%	561	100%
	Div.	434	61%	2	0.3%	158	22%	114	16%	708	100%
COMMODITY	Acq.	351	61%	42	7%	132	23%	47	8%	572	100%
	Div.	277	78%	4	1%	60	17%	15	4%	356	100%
LIFE SCIENCE	Acq.	362	49%	40	5%	233	31%	109	15%	744	100%
	Div.	225	51%	15	3%	118	27%	84	19%	442	100%
OTHER CHEMICAL	Acq.	509	45%	66	6%	277	25%	274	24%	1,126	100%
	Div.	326	51%	13	2%	121	19%	176	28%	636	100%
OTHER INDUSTRIES	Acq.	250	40%	13	2%	177	29%	180	29%	620	100%
	Div.	188	46%	1	0.2%	127	31.3%	90	22%	406	100%
TOTAL	Acq.	1,798	49.9%	166	4.6%	967	26.7%	692	19.1%	3,623	100.0%
	Div.	1,450	56.9%	35	1.4%	584	22.9%	479	18.8%	2,548	100.0%

** Continental Europe.*

Table 3a: R&D intensity and N. of firms with R&D data, by segment and year

	1987		1988		1989		1999		1991		1992		1993		1994		1995		1996		1997	
	RDI	N	RDI	N	RDI	N	RDI	N	RDI	N	RDI	N	RDI	N	RDI	N	RDI	N	RDI	N	RDI	N
ENERGY	0,70%	35	0,70%	36	0,80%	42	0,80%	42	0,90%	42	0,90%	42	0,90%	46	0,80%	45	0,70%	48	0,60%	42	0,70%	39
COMMODITY	3,10%	34	3,10%	45	3,30%	54	3,40%	58	3,50%	61	3,60%	64	3,80%	66	3,50%	63	3,10%	79	3,10%	82	3,80%	71
LIFE SCIENCE	8,80%	79	9,00%	101	8,60%	116	8,60%	118	9,00%	119	9,60%	120	10,10%	121	9,80%	124	10,40%	143	10,90%	141	11,00%	113
OTHER CHEMICAL	3,70%	60	3,50%	65	3,80%	81	3,90%	83	3,90%	89	4,10%	90	4,00%	93	3,60%	91	3,20%	106	3,20%	109	3,20%	93
OTHER INDUSTRIES	3,50%	27	3,10%	31	3,00%	40	3,50%	41	3,70%	41	3,80%	43	3,50%	44	3,40%	44	3,10%	45	2,90%	45	3,10%	42
Total	2,60%	235	2,80%	278	2,90%	333	2,90%	342	3,20%	352	3,50%	359	3,60%	370	3,60%	367	3,40%	421	3,40%	419	3,30%	358

Table 3b: Average R&D intensity 1987-97 , by segment and region

	North America	Japan	EU*	UK	Total
ENERGY	0.7%	0.6%	0.9%	0.7%	0.8%
COMMODITY	3.3%	3.3%	3.5%	2.3%	3.3%
LIFE SCIENCE	13.1%	9.4%	8.7%	9.0%	10.7%
OTHER CHEMICAL	3.3%	3.6%	3.4%	2.5%	3.3%
OTHER INDUSTRIES	3.2%	2.8%	3.6%	1.4%	3.0%
Total	3.3%	4.0%	3.1%	2.7%	3.2%

While the small number of Japanese deals is probably due to a reporting bias in the two databases we use, the relatively larger volume of U.S. and UK transactions is likely to reflect a genuine phenomenon. Indeed, Arora and Gambardella (1998) found that Japanese acquisitions in chemicals between 1985-1993 were less than 10% of those by U.S. chemical firms. Moreover, over 70% of the Japanese acquisitions were overseas, in contrast to 57% by European firms and only 12% by U.S. firms. Table 2 reveals that our sample firms have more acquisitions than divestitures, except in the Energy segment. This suggests that some private firms were divesting divisions to our sample firms or, more likely, that firms outside the chemical sector were divesting their chemical businesses, as for instance was true of Allied Signal or U.S. Steel. The second trend that emerges from table 2 is that U.S. firms were more likely than their European or Japanese peers to be divesting – the ratio of divestitures to acquisitions is higher in the U.S. than elsewhere, even within segments. Put differently, although U.S. firms account for the lion's share of all restructuring in each segment, they account for a greater share of the divestitures than of acquisitions. By contrast, Japanese and European firms are much more likely to be acquirers, relatively speaking. British firms are somewhere in between. This pattern is consistent with the notion that equity market pressures on management to cut costs and focus are more prominent in the U.S., and to a lesser extent, in the UK. Table 3 documents the time pattern of R&D intensity by segment. R&D intensity is stable over time. Table 3b shows the importance of small firms in the U.S. life sciences segment. These firms tend to have high R&D intensity, principally because their sales are small. Other than that, the different regions are similar in terms of R&D intensity within segments, with any differences across regions being due to differences in the relative weights of segments. Thus, Japan has a higher R&D intensity, in large measure because it is relatively intensive in life sciences and specialty chemicals.

3. R&D AND CORPORATE RESTRUCTURING

3.1 A Model of R&D Expenditure

We assume that firms operate in many segments, each with a different level of R&D intensity. Further, we assume that R&D intensity may vary by scale. Thus we write:

$$R_{ij} = \alpha_j X_{ij}^{\beta}$$

where: R is R&D, X is sales, *i* denotes firm, *j* denotes industry segment, β is the scale parameter, which reflects the elasticity of R&D with respect to size, and α_j is a parameter that measures the R&D intensity of a unit-sized firm in segment *j*. For simplicity, we set the size of the median-sized firm to unity, so that α_j measures the R&D intensity of the median firm in segment j. We denote by S_{ij} the revenue share of the *j*-th industry segment for firm *i*, and write:

$$R_i = \Sigma_j R_{ij} = \Sigma_j \alpha_j X_{ij}^{\beta} = X_i^{\beta} \Sigma_j \alpha_j S_{ij}^{\beta}$$

Taking logs, letting lower case represent logged values, and using time subscripts, yields the following expression:

$$r_{it} = \beta x_{it} + log(\Sigma_j \alpha_{jt} S_{ijt}^{\beta}) \quad (1)$$

Note that the second term, which can be interpreted as the scale-adjusted R&D intensity, is not observed. In fact, only U.S. listed companies are required to report sales by business line, but even here business lines typically do not fit with a technologically meaningful partition of the industry. Further, S_{ij} is likely to vary across firms and within a firm, due to restructuring and to changes in segment level R&D intensity.

This framework implies that restructuring affects R&D in two ways: *(i)* through a change in scale, and *(ii)* through a change in scale-adjusted R&D intensity. In other words, R&D may be affected both by changes in scale and by segment specific shifts in scale-adjusted R&D, since α_j may vary with time. Moreover, there could conceivably be firm-specific shifts in R&D. We can allow for such firm-specific effects by specifying the R&D equation as:

$$R_i = C_i\Sigma_j R_{ij} = C_i\Sigma_j \alpha_j X_{ij}^{\beta} = C_i X_i^{\beta}\Sigma_j \alpha_j S_{ijt}^{\beta} \qquad (1a)$$

where C_i represents these firm specific effects. We face two basic problems in order to estimate equation (1) or (1a): the unobservability of S_{ijt} and the non-linearity in the specification. A solution to the first problem comes from the fact that we can observe the number of acquisitions and divestitures, which represent a change in S_{ijt}, but even here we face the problem that S_{ijt} being a revenue share, it increases with the number of (net) acquisitions and with the sales associated with each transaction, and it decreases with the existing scale of the firm—i.e. one acquisition will imply a bigger change in revenue shares for a smaller firm as it will for a larger firm. In order to tackle the non-linearity, we write:

$$\log(\Sigma_j \alpha_{jt} S_{ijt}^{\beta}) = A_i + W_t + B_j + \Sigma_j \gamma_j z_{ijt}$$

where A_i, W_t, and B_j are firm, time and industry segment specific effects, and z_{ijt} represents the number of net *scale-adjusted* acquisitions in industry segment j by firm i at time t. This suggests an estimation strategy which relies on estimating:

$$r_{it} = \beta x_{it} + A_i + W_t + B_j + \Sigma_j \gamma_j z_{ijt} \qquad (2)$$

Here we assert that changes in R&D intensity across segments are common over time and captured by W_t—i.e. there are no time and segment specific effects, only time specific effects and segment specific effects. The segment specific effects are captured by B_j, and that the unobserved initial scale and composition effects are captured by A_i, which are unobserved firm-specific fixed effects. We implement this approach by estimating the following equation, which represents our preferred specification:

$$\ln(R\&D)_{it} = \alpha + \beta\ln(Sales_{it-1}) + \gamma_1\ln(CF_{it-1}) + \gamma_2\ln(LR_{it-1}) + \gamma_3\ln(D_{it-1}) + \gamma_4\ln(E_{it-1}) + \delta_1 Aen_{it-1} + \delta_2 Acn_{it-1} + \delta_3 Aln_{it-1} + \delta_4 Aotn_{it-1} + \alpha_i + \eta_t + \varepsilon_{it} \quad (3)$$

where:

R&D is given by all costs incurred during the year that relate to the development of new products or services;

Sales is gross sales reduced by cash discounts, trade discounts, and returned sales and allowances for which credit is given to customers;[17]

CF (Cash Flow) is defined as Income Before Extraordinary Items, i.e. a company's income after all expenses except provisions for dividends, to which depreciation and amortization charges are added;

LR (Liquidity Ratio) is the ratio of total current assets to total current liabilities. Total current assets represents cash and assets expected to be realized in cash within next year. Total current liabilities represents debt and other liabilities due within one year;

D is debt, i.e., the sum of all corporate debts, irrespective of maturity;

E represents total shareholders' equity;

Aen represents net acquisitions (acquisitions net of divestitures) in energy related industries;

Acn represents net acquisitions in commodity chemicals;

Aln represents net acquisitions in life science industries;

Aotn represents net acquisitions in other industries, chemical or not.[18]

[17] We also estimated with current period sales, and with both current and lagged sales. The qualitative results are unchanged.

[18] We also run our regressions using the two segments ('other chemical' and 'other (non chemical)' industries) separately. Our results are confirmed, but at the price of losing a large number of observations. The reason is that restructuring deals are coded by Securities Data Company using both the SIC and the Venture Economics industry classifications. The latter has the advantage of being much more detailed for high-technology than for other industries. As a result, however, many deals have a missing Venture Economics industry code that we recoded as 'other industries.' We thus employ the SIC code associated with each SDC transaction.

We include four financial measures, cash flow, the liquidity ratio, debt, and equity. We do so because the literature on investment equations suggests that the most likely candidate for firm-specific effects in decisions on intangible investments are financial variables, to the extent that they reflect the severity of firm-specific financial constraints. We thus need to control for such effects when assessing how restructuring affects R&D investment. There are several reasons why firms may face credit constraints and may thus forego attractive investment opportunities, as argued in the large literature which developed since the seminal work of Fazzari, Hubbard and Petersen (1988), and which is surveyed in Hubbard (1998) Kaplan and Zingales (1997), and Schiantarelli (1996).

Cash flow is relevant because it may shield a firm from the need to look for external finance. Hence we expect γ_1 to be positive.[19] The liquidity ratio gives a measure of how a firm may use working capital to absorb possible credit crunches, given the high cost of changing the R&D investment. We thus expect γ_2 to be positive. Debt and equity reflect a firm's capital structure. We expect γ_3 to be negative since more debt absorbs cash flow to be serviced. On the contrary, we expect γ_4 to be positive, since equity provides a more suitable form of financing for uncertain, long-term forms of investment like R&D.

All the regressors are lagged one year. Given the relatively small time span (eleven years), we estimate η_t by including a full set of year dummies in all the specifications. We estimate this specification with both random effects and fixed effects. As we will see shortly, the random effect model is generally rejected, so that we base our main results on the fixed effect estimates. Our analysis is based on counts of transactions, acquisitions and divestitures. We also have data on the report value of the transaction for about 50% of the deals. However, these data are not randomly distributed across firms; in particular, transactions involving European and Japanese are more likely to lack values. Moreover, we are concerned about the validity of the reported deal values, since these may not be assigned in a consistent way. Nonetheless, although we do not base ourselves on estimates involving values, we estimate a specification using the value of net acquisitions, reaching similar results.

[19] It is common in the literature to control for investment opportunities using Tobin's q as a regressor, though this is a debated practice. This information is not available for Continental European and Japanese firms. We control for segment specific opportunities to some extent by running within segment regressions, but, since we use a firm fixed effect specification, we are unable to account for firm specific time varying opportunities.

We also test a more general specification with acquisitions and divestitures included separately. Finally, we test a specification with the aggregate number of transactions to proxy for the level of restructuring activity. The results of these specifications are consistent with our preferred specification.

Table 4 provides descriptive statistics for our panel. The average R&D expenditure is \$135 millions (median \$21 millions). Notice that nearly a third of the observations are missing. The average sales in our sample is about \$4 billions (median \$800 million). On average, sample firms have divested more than acquired in Energy (the average number of net acquisitions in energy related industries is -0.03), and acquired more than divested in Other Chemicals and in Other (non chemical) industries (the average number of net acquisitions in other industries is .12). This is consistent with the idea that more firms with substantial chemical businesses were trying to build up their non-chemical businesses. Within the chemical industry firms have the largest number of net acquisitions in Life Sciences (0.03). The average number of acquisitions and divestitures in the sample—not reported in the Table—is approximately 7 and 5, respectively.

Table 4 : Pooled sample descriptive statistics (535 firms, 11 year period)

Variable	Obs.	Mean	Median	Stand. Dev.	Min	Max
R&D^{+}	3 884	135	21	304	0	2 961
Log(R&D)$^{+}$	3 800	3,13	3,09	2,09	-4,71	7,99
Sales^{+}	4 370	4 151	805	10 436	0	128 207
Cash flow^{+}	4 089	481	69	1 287	-1855	18 717
Debt^{+}	4 372	1 274	187	5 875	0	162 406
Equity^{+}	4 087	1 716	416	4 312	-1563	63 620
Liquidity Ratio	4 347	2,74	1,74	3,6	0	57,26
Net No. of Acq. - Energy	5 885	-0,023	0	0,466	-10	5
Net No. of Acq. - Commodity	5 885	0,016	0	0,487	-6	5
Net No. of Acq. - Life Sc.	5 885	0,031	0	0,446	-4	5
Net No. of Acq. - Other Ind.	5 885	0,117	0	0,792	-7	9

Note: (1) Data for the 535 firms and the 11 years in the sample (unbalanced panel). (2) + indicates data in millions of US current dollars. For the empirical analysis we used the natural logarithm of these variables. (3) Net acquisitions represent the number of acquisitions minus the number of divestitures.

3.2 The Impact of Restructuring on Firm-Level R&D

In Table 5 we report panel data fixed effects ("within") estimates of (2) for the pooled sample, by segment, and by world region. There are some noteworthy results which emerge from this analysis. Restructuring matters, with net acquisitions in different industries having an effect on R&D, even controlling for size. If one looks at the pooled regression, we find that the estimated value of β is 0.58. The sign of the coefficients of the restructuring variables are as expected, with net acquisition in the R&D intensive segment—in particular Life Science—having a positive and significant impact on R&D, even after controlling for size, consistent with the interpretation in (2). The coefficients of the financial variables are all positive, but only the coefficient of equity is significant at conventional significance levels.

We performed our regression analysis using both the net number of acquisitions and the *cumulative* value (unreported), i.e. the number of acquisitions since 1987 cumulated through the end of each observation year. On the whole the two specifications yield very similar qualitative results, although in specifications using net cumulative acquisitions, the estimated elasticities tend to be slightly lower.

We also estimate (2) using a random effects model, which is rejected, using a Hausman specification test, for the aggregate case and for Life Science, but cannot be rejected for the Other Industries segment. At the regional level, the random effects is rejected for North America and Japan, but not for Europe and UK. For the North America case, the random effects model is rejected for the Life Science segment and Other non chemical industries only. These tests indicate that there are both country and industry specificities affecting the relationships studied, suggesting to run the analysis within both levels. Within segment estimates are potentially useful also because restructuring events may not be distributed randomly across firms. For instance, Life Science firms may be more likely to be involved in life science related acquisitions. They also better control for differences in technological opportunities across segments.

By estimating (2) within industry segment, we allow the time effects to vary by industry segment and we also allow β to vary by industry segment. The within industry segment specification is useful for another reason: it allows the coefficients on z_{ijt}, γ_j to vary by industry segment as well, which is desirable because the second term depends on β, and if β varies by industry segment, so should the coefficients. Further, since the acquisition of a commodity unit by a pharmaceutical firm should have a different effect on scale adjusted R&D intensity than the acquisition of the same unit by an energy firm, we should allow the coefficient of net acquisitions to vary by the industry segment of the restructuring firm. Clearly, the usefulness of this procedure depends on the accuracy of the classification of firms by industry segment.

Table 5 : Fixed Effects estimation results, pooled sample, within segment and within world region

Dependent Variable: Firm level Log(R&D)

	Pooled	Energy	Commodity	Life Science	Other Chemical	Other Industries	North America	Japan	EU*	UK
Log of sales	0.58	0.42	0.27	0.57	0.74	-0.10	0.50	0.18	1.60	0.99
	(0.05)	(0.20)	(0.12)	(0.12)	(0.07)	(0.15)	(0.05)	(0.20)	(0.36)	(0.19)
Log of Cash flow	0.04	0.10	0.10	-0.01	0.04	0.06	0.05	-0.11	-0.08	0.11
	(0.02)	(0.05)	(0.05)	(0.05)	(0.02)	(0.05)	(0.02)	(0.08)	(0.16)	(0.06)
Liquidity Ratio	0.05	-0.25	-0.04	0.16	0.07	-0.16	0.06	-0.07	0.22	-0.39
	(0.04)	(0.11)	(0.08)	(0.09)	(0.06)	(0.11)	(0.04)	(0.16)	(0.22)	(0.18)
Log of Debt	0.03	0.07	0.07	-0.01	0.04	0.10	0.02	-0.04	-0.07	0.09
	(0.01)	(0.05)	(0.03)	(0.03)	(0.02)	(0.05)	(0.01)	(0.04)	(0.09)	(0.06)
Log of Equity	0.16	0.00	0.12	0.06	0.15	0.31	0.19	0.50	-0.12	0.12
	(0.04)	(0.13)	(0.07)	(0.10)	(0.04)	(0.08)	(0.03)	(0.18)	(0.22)	(0.10)
Net Acq. Energy	0.01	0.03	0.00	0.00	0.00	0.03	0.01	-0.05	0.03	-0.02
	(0.01)	(0.03)	(0.02)	(0.09)	(0.01)	(0.04)	(0.02)	(0.08)	(0.04)	(0.03)
Net Acq. Commodity	0.01	0.00	0.04	..	0.04	0.06	0.02	..	-0.03	-0.03
	(0.02)	(0.02)	(0.04)	..	(0.06)	(0.10)	(0.01)	..	(0.06)	(0.04)
Net Acq. Life Science	0.04	0.04	0.05	0.05	0.03	0.05	0.05	0.09	0.01	0.05
	(0.02)	(0.17)	(0.04)	(0.03)	(0.03)	(0.04)	(0.02)	(0.10)	(0.05)	(0.04)
Net Acq. Other Ind.	0.01	0.03	0.03	-0.01	0.01	0.00	0.02	0.01	0.02	0.00
	(0.01)	(0.02)	(0.02)	(0.04)	(0.01)	(0.02)	(0.01)	(0.06)	(0.04)	(0.02)

*Notes: Standard errors in parenthesis; *: Continental Europe.*

There is considerable variation in the within industry segment estimates, in both the elasticity of R&D with respect to scale and in the impact of restructuring. The estimated elasticity of R&D with respect to sales varies between 0.27 (Commodities) and 0.74 (Others, Chemical and Non-chemical). As far as the impact of restructuring is concerned, net acquisitions in Life Sciences, and to a lesser extent, in Other Chemicals, tend to increase R&D, particularly in less R&D intensive segments like commodities. Overall, this pattern is consistent with the idea that the addition of a business to a firm's R&D portfolio with R&D intensity higher than the average R&D intensity of the portfolio will increase the average R&D intensity of the portfolio. However, since an acquisition is likely to bring about scale changes, and since scale changes may affect R&D differently in different segments—so that contrary to the maintained assumption, β may differ across segments— the economic meaning of the coefficients of the net acquisition variables cannot be directly interpreted. Note also that the significance of the restructuring variables tends to be lower in the within industry segment estimates. This drop in significance is due to the reduced number of observations—recall that most of the key variables are identified by within firm variation over time, rather than across firms.[20]

Note also that variation in estimated coefficients is present even if we look at regional-level regressions: β becomes small and insignificant for Japan, and larger than one (1.6), and significant, for Europe (and close to one for the UK). The segment-level estimates for North America are however very close to those for the pooled sample.

We also examine the direct impact of financials on R&D. At the aggregate-pooled level we find that all financial variables have positive coefficients, although the elasticities are low, and vary across segments. In the pooled estimates, the coefficient estimates of debt and equity imply that a capital structure tilted towards debt restrains R&D, as we would expect. It also appears that firms with a Life Sciences focus are different from other chemical firms in terms of the relative unimportance of financial variables. For instance, cash flow is negative and insignificant as is debt, and the other two, liquidity and equity are positive but insignificant. Interestingly, financial variables turn out not to be

[20] The foregoing specifications implicitly assume that R&D changes are a function only of the restructuring of the past year. This assumption is questionable because R&D tends to be a long term investment. It is further questionable because theory suggests that the impact of restructuring should be long lived, and one-year-lagged sales may not adequately account for this. One way of modeling this is to think of R&D as adjusting over time to a restructuring event. The simplest model of that process implies that lagged R&D is included as an explanatory variable. Such specification is often used in order to take care of adjustment costs and expectations about the future profitability of investments. In an exploratory exercise (unreported) we then estimate a modified version of (2) which includes the endogenous 1-year lagged values of the natural logarithm of R&D on the right-hand-side using a fixed effects estimator and a two stage least square procedure, with the 2-year lagged R&D as instrument. We obtain similar, somewhat smaller values for β, which is 0.61 for the whole sample and ranges from 0.38 for Commodities and 0.65 for Life Sciences, and is always significant. The lower values of the segment-level estimates suggests that firms may not adjust their R&D to annual changes in scale, but instead the adjustment period may be longer than one year, and that it may be particularly long for drugs but shorter for other industries.

significant for Continental Europe and Japan, two regions where the structure of the financial systems is often thought to be conducive to looser financial constraints.[21]

We also find that financial variables do not affect the impact of restructuring on R&D. In other words, whether we include financial ratios or not in the specification, the coefficient of net acquisitions do not vary much, both in the pooled and in the segment level estimates. This is an important result, since it means that restructuring is not affecting R&D solely through its impact on the firm's financial position, for instance by increasing or decreasing liquidity or debt, as it may have appeared from the results of Bronwyn Hall's previous studies.

An important assumption of the econometric framework thus far is that only net acquisitions matter, i.e., the coefficient on acquisition is equal and opposite in sign to that on divestiture. This is a testable proposition, but our experiments on this are inconclusive. One reason might be multicollinearity, another that acquisitions and divestitures are not random. Indeed, in some cases, acquisitions may be positively correlated with technological opportunities available to the firm in other segments, while in other cases they may be negatively correlated. To explore this further, we divided all firms into active and inactive with respect to their restructuring behavior. We define a firm as active (inactive) if it had more (less) than the median number of transactions—which is three for the sample as a whole. Further, active firms were classified into net acquirers, net divestors and balanced. We then compared R&D in two periods, 1990-91 and 1996-97 using two-year averages by the industry segment. The results are shown in Table 6.[22] The first point to note is that active acquirers tend to be more R&D intensive in most segments. Since active firms are unlikely to be the small firms (see below), this implies that the higher R&D intensity is not due to scale effects but measures genuine differences in portfolio composition. Among active firms, acquirers are more R&D intensive than divestors (and increasingly so over time) in Energy, Commodity and Other Chemicals, whereas the opposite is true for Life Sciences and Other non-chemicals. Another point to notice is that active divestors in Commodities see their R&D intensity grow over time. There does not appear to be any clear trend in terms of changes in R&D intensity over time for the sample as a whole.

[21] Bond, Harhoff and Van Reenen (1999) specifically test the sensitivity of R&D expenditure to financial constraints in Germany and Britain, and find analogous results using a variety of specifications.

[22] The very high R&D intensity for Life Science is due to the presence of a number of small biotechnology firms.

Table 7, finally, shows the acquisitions and divestitures by industry segment, region, and firm size. It suggest that restructuring is overall related to size, and also that firms active in restructuring tend to be more R&D intensive. These figures suggest that in R&D intensive industry segments, R&D intensive firms are more likely to be acquirers, while the reverse is true for the less R&D intensive segments.

Table 6 : average R&D intensity by segment and acquisition behavior, 1990-91 and 1996-97

		Inactive			Active		
		Balanced	Acquirers	Divestors	Balanced	Acquirers	Divestors
ENERGY	(1)	0.7%	0.4%	1.3%	1.9%	2.1%	0.7%
	(2)	0.5%	0.8%	--	2.3%	1.6%	0.4%
COMMODITY	(1)	1.7%	4.1%	5.2%	4.6%	3.8%	2.4%
	(2)	2.5%	3.3%	4.1%	4.5%	3.8%	3.0%
LIFE	(1)	8.7%	8.7%	7.8%	41.3%	8.6%	10.2%
SCIENCE	(2)	11.0%	10.8%	9.0%	46.7%	10.6%	12.9%
OTHER	(1)	3.4%	1.2%	1.6%	3.0%	4.1%	4.3%
CHEMICALS	(2)	2.4%	1.7%	3.9%	2.7%	3.5%	2.1%
OTHERS	(1)	3.0%	2.8%	3.4%	2.8%	3.4%	5.7%
	(2)	2.3%	2.6%	7.5%	2.8%	2.6%	3.5%
TOTAL	(1)	3.0%	4.0%	2.3%	2.5%	4.7%	1.4%
	(2)	3.1%	4.1%	6.2%	3.0%	4.8%	1.4%

(1) = R&D intensity in 1990-91
(2) = R&D intensity in 1996-97

Table 7 : Restructuring volumes by industry, region, and firm size.

		Small firms					Large firms				
		North America	Japan	EU*	UK	Total	North America	Japan	EU*	UK	Total
ENERGY	Acq.	48	1	7	10	66	278	4	141	72	495
	Div.	38	0	3	10	51	396	2	155	104	657
COMMODITY	Acq.	54	0	10	6	70	297	42	122	41	502
	Div.	26	0	3	4	33	251	4	57	11	323
LIFE SCIENCE	Acq.	167	5	13	53	238	195	35	220	56	506
	Div.	134	4	5	11	154	91	11	113	73	288
OTHER CHEMICALS	Acq.	156	5	8	69	238	353	61	269	205	888
	Div.	111	0	3	17	131	215	13	118	159	505
OTHER INDUSTRIES	Acq.	47	0		10	57	203	13	177	170	563
	Div.	32	0		11	43	156	1	127	79	363
Total	Acq.	472	11	38	148	669	1,326	155	929	544	2,954
	Div.	341	4	14	53	412	1,109	31	570	426	2,136

Note: Large firms are those with sales greater than the median value of sales for the sample period ($ 682 million).
**: Continental Europe.*

4. CONCLUSIONS

Restructuring can affect R&D intensity in a variety of possible ways. The most obvious way is by leading managements to focus on short run performance to the possible detriment of the long run performance. Typically, this happens by changing the financial condition of the firm, for instance by changes in debt or liquidity. Acquisitions financed by debt, for instance, are thought to reduce R&D spending. However, there are at least two other ways which our study captures. First, R&D investments may be subject to scale effects, and acquisitions increase scale and thereby might stimulate R&D. Second, R&D intensity may differ across industries and industry segments, and hence, the type of acquisition or divestiture should also matter for R&D, controlling for scale.

To the best of our knowledge, our study is the first to explore firm-level data for an analysis of the effects of restructuring on R&D using data on both acquisitions and divestitures, and a panel composed of a large number of chemical companies from many countries.

We focus on a single industry, chemicals, which has undergone a long period of thorough restructuring, and for which we are able to identify technologically homogeneous segments. We find that restructuring matters for R&D, and that (net) acquisitions of businesses in research-intensive segments have the highest effect. The considerable variation in segment-level estimates documents the different nature of the dependence of R&D on restructuring across industry segments. We also find that financial variables, while affecting R&D as already known, do not affect the way restructuring affects R&D. This finding suggests that omitting the type of acquisition or divestiture and failure to control for scale may distort the estimated effect of restructuring on R&D.

5. REFERENCES

Aftalion, Fred (1989), History of the International Chemical Industry, Philadelphia, University of Pennsylvania Press.

Arora, Ashish, and Alfonso Gambardella (1998) 'Evolution of Industry Structure in the Chemical Industry,' in Ashish Arora, Ralph Landau, and Nathan Rosenberg (eds), Chemicals and Long-term Growth, New York, Wiley & Sons.

Arora, Ashish, Landau Ralph, and Nathan Rosenberg, (eds.) (1998), Chemicals and Long-term Economic Growth, New York, John Wiley & Sons.

Ahuja, Gautam, and Riitta Katila (2001), 'Technological Acquisitions and the Innovation Performance of Acquiring Firms: A Longitudinal Study,' Strategic Management Journal, v.22, p.197-220.

Blair, Margaret, and Robert Litan (1990), "Corporate Leverage and Leveraged Buy-Outs in the Eighties," in John Shoven and Joel Waldfogel (eds.) Debt, Taxes, and Corporate Restructuring, Washington, DC, Brookings Institution, p.43-80.

Blair, Margaret, and Martha Schary (1993) 'Industry-level Pressures to Restructure,' in Margaret Blair (ed.) The Deal Decade, Washington, D.C., The Brookings Institution.

Bond, Steven, Harhoff Dietmar, and John Van Reenen (1999) 'Investment, R&D and Financial Constraints in Britain and Germany,' Institute of Fiscal Studies Working Paper 99-05.

Cabral, Antonio (2000), Introduction to Industrial Organization, Cambridge, MIT Press.

Chandler, Alfred (1990), "Scale and Scope: the Dynamics of Industrial Capitalism," Cambridge, Harvard University Press.

Cohen, Wesley (1995), "Empirical Studies of Innovative Activity," in Paul Stoneman, ed., Handbook of the Economics of Innovation and Technological Change, Oxford, Basil Blackwell.

Fazzari, Steven, Hubbard, Robert and Bruce Petersen (1988), 'Financing Constraints and Corporate Investment,' Brooking Papers on Economic Activity: Microeconomics, Washington D.C., The Brookings Institution, p.141-95.

Fusfeld, Herbert (1987), 'Corporate Restructuing—what impact on U.S. industrial research?' Research Management, v.30 (July-AUgust), p.10-17.

Hall, Bronwyn (1988) "The Effect of Takeover Activity on Corporate Research and Development," in Alar Auerbach (ed.) Corporate Takeovers: Causes and Consequences, Chicago, University of Chicago Press, 69-96.

Hall, Bronwyn (1990) "The Impact of Corporate Restructuring on Industrial Research and Development," in Brookings Papers on Economic Activity: Microeconomics, Washington D.C., The Brookings Institution, p.85-135.

Hall, Bronwyn (1992) "Investment and R&D at the Firm Level: Does the Source of Financing Matter?" NBER Working Paper 4096.

Hall, Bronwyn (1994) "Corporate Restructuring and Investment Horizon in the United States, 1976-1987," Business History Review, Spring 1994, p.110-43.

Hall, Bronwyn (1999), "Mergers and R&D Revisited," mimeo, University of California at Berkeley.

Hall, Bronwyn, Mairesse Jaques, Branstetter, Leo, and Bruno Crepon (1999) 'Does Cash Flow Cause Investment and R&D? An Exploration Using Panel Data for French, Japanese and United States Scientific Firms,' in David Audretsch and A.R. Thurik (eds.) Innovation, Industry Evolution and Employment, Cambridge, Cambridge University Press.

Hoskisson, Robert, and Richard Johnson (1992), 'Corporate Restructuring and Strategic Change: the Effect on Diversification Strategy and R&D Intensity,' Strategic Management Journal, v.13, p. 625-634.

Hubbard, Robert (1998) 'Capital-Market Imperfections and Investment,' Journal of Economic Literature, v.36, p.193-225.

Kaplan, Stephen (1989) "The Effects of Management Buyouts on Operating Performance and Value," Journal of Financial Economics, v.24, p.217-54.

Kaplan, Stephen, and Luigi Zingales (1997), 'Do Investment-Cashflow Sensitivities Provide Measures of Financing Constraints?' Quarterly Journal of Economics, .v.112, n.1, 169-215.

Lichtenberg, Frank, and Donald Siegel (1990) "The Effects of Leveraged Buyouts on Productivity and Related Aspects of Firm Behavior," Journal of Financial Economics, v.27, p.165-94.

Miller, Robert (1990), 'Do Mergers and Acquisitions Hurt R&D?' Research Management, v.33 (March-April), p.11-15.

Martin, Stephen (2001), Advanced Industrial Economics, Second Edition, Blackwell, Oxford.

Mitchell, Mark, and Hardold Mulherin (1996), 'The impact of industry shocks on takeover and restructuring activity,' Journal of Financial Economics, v41, n1, 193-229.

Schiantarelli, Fabio (1996), 'Financial Constraints and Investment: A Critical Survey of International Evidence,' Oxford Review of Economic Policy, v.12.

Smith, Abbie (1990), "Corporate Ownership Structure and Performance: The Case of Management Buyouts, " Journal of Financial Economics, v.27, p.127-64.

Stein, Jeremy (1988) "Takeover Threats and Managerial Myopia," Journal of Political Economy, v.96, p.61-80.

Tirole, J. (1988), The Theory of Industrial Organization, Cambridge, MA, The MIT Press.

PART 2:

TECHNOLOGY DEVELOPMENT AND DIFFUSION

Chapter 4

WATCHING PAINT DRY?

Regulation, Demand and Innovation in a Mature Industry

Stefano Brusoni
CESPRI,CRORA and Bocconi University

Abstract: The development and diffusion of environmentally safe products is a key political concern in industries potentially responsible of polluting productions, like chemicals in general, and paints and coatings in particular. On the basis of ten interviews carried out with industry experts and practitioners, this chapter looks at the innovative dynamics of the European paints industry, and explores the role of environmental regulation and customers' behaviour in inducing the development and diffusion of clean technologies and products. Specifically, the chapter focuses on the regulation concerning the emissions of Volatile Organic Compounds (VOCs), and the consequent introduction of water-based paints. One of the main results is that, from a technological point of view, the European industry seems to be perfectly capable of coping with the diffusion of new, solvent free formulations. No difference seems to exist between US and European firms. Rather than a lack of technological capabilities, customer demand is the key to some of the problems related to the diffusion of water-based products. Concerning the role of environmental regulation, stringent regulations are not enough to induce the diffusion of cleaner technologies.

Key words: Environmental Regulation; Innovation Incentives; Paints and Coatings Industry

1. INTRODUCTION

This chapter focuses on the recent evolution of the EU paints and coatings industry, with respect to the diffusion (or lack thereof) of clean technologies and products. The main argument of this chapter is that, in order to understand the innovative dynamics of this industry, one has to look at the interplay of environmental regulation on the one side, and customers' behavior on the other.

F. Cesaroni et al. (eds.), R&D, Innovation and Competitiveness in the European Chemical Industry, 93-118.

It is argued that stringent regulations are not enough to induce the diffusion of cleaner technologies. In mature industries such as the one studied in this chapter, the willingness of customers to switch to new types of products, as well as producers' access to distribution channels, play a fundamental role in understanding the diffusion of new, cleaner technologies and products.

Throughout the 1990s, environmental regulations (actual or threatened) have acted as a major driving force in activating R&D efforts in an otherwise rather conservative sector. In this respect, this chapter will focus on the role played by regulatory efforts concerning the emissions of Volatile Organic Compounds (VOCs) during manufacturing and application. VOCs are compounds that vaporize during an industrial process. Such a broad definition includes a rather heterogeneous set of chemicals whose effects have a variety of different environmental consequences. These include stratospheric ozone depletion, greenhouse effect, formation of ground level ozone, and odor nuisance. Some compounds are deemed to be toxic and carcinogenic. This chapter focuses on VOCs for two reasons. First, in 1999 the EU issued a comprehensive directive aimed at dealing with VOC emissions problems. During the 1990s, the debate that accompanied the preparation of this directive acted as 'focusing device' for the industry. VOC emissions control became a key issue throughout this composite sector. Second, while regulatory pressures and the related debate impacted on the entire industry, different segments reacted in different manners. Some were more pro-active than others. In particular, it seems that the industrial coatings end of the business was very quick in responding (preventively) to regulation, adopting clean(er) technologies capable of vastly reducing VOCs emissions. The decorative paints business took longer to react. This chapter focuses on the latter, using the former as a 'control case'. It argues that delays in adopting VOC-free products in the decorative paints business were not related to the lack of the capabilities needed to develop such products, but rather to customers' conservative attitudes and lack of access to distribution channels.

On the methodological side, the information reported in the following sections is based upon both primary and secondary sources. On the basis of a thorough review of the relevant literature (both academic and trade publications), ten interviews were carried out involving both industry experts and practitioners. The experts comprise four people working for government and industry associations who were consulted to shape and then put into a wider context the interviews conducted. Three of these where 'segment experts' (respectively coatings, dyes and pigments, and printing inks); the fourth is the environment officer of a leading sector association. In the decorative paints business, five people were interviewed. They all work for a leading manufacturer and have senior positions in R&D; marketing; technology and safety; health and environment management; and manufacturing.

This chapter is structured as follows. Sections 2 and 3 describe the industry, its key market segments, and their competitive dynamics. Section 4 discusses the drivers and direction of technical change, while section 5 focuses on the relationship between environmental regulation and innovation. Section 6 analyses the problems generated by high VOC emissions, and various alternatives to curb them. Section 7 compares the diffusion of waterborne (VOC-free) paints in two lines of business: decorative and motor vehicle paints. Finally, section 8 will conclude.

2. SOME DEFINITIONS

The term paint is usually meant to include all types of surface coatings that are used for decorating and protecting substrates as well as for various special applications. Coatings are complex chemical compounds involving the blending of several organic and inorganic components in varying proportions. The ingredients reported in any 'coatings cookbook' are resins (also vehicles or binders), pigments, solvents (also surfactants) and numerous additives. The vehicle contains the pigment (usually in the form of a powder providing the coloring and obscuring properties). Solvents are added to improve the spreading properties of the vehicle. Solvents are usually based on organic liquids (hence the problem with VOC emissions), though water is being used to an increasing extent. Finally, additives, though not essential, are usually employed to improve the properties of the compound and facilitate its application (e.g. biocides).

Resins are by far the most critical component. They determine both the overall production costs and the final product performance. Therefore, the largest paint producers tend to manufacture this component in house, while buying in pigments, additives and solvents. Changes in paint formulation usually involve a change in the formulation of the resin. It is unlikely that changes in the formulation of other components will bring about a major change at product level. Changes in the resin, on the other hand, are more likely to have far reaching implications for the final product. It is indeed possible to reduce VOC emissions without switching to waterbased paints, i.e. without changing the solvent employed. However, this requires that the other components be adapted to achieve lower emission levels. Also, the cost increases and the performance is likely to worsen, particularly in terms of color delivery over time.

Classifying paints is a risky business. Several typologies are available based upon alternative criteria: end-use, solvent system, solids content, type of curing, etc. Basically, the criterion adopted varies according to the aim. One of the most commonly adopted is based upon the resin used. The most widely used resins are polyesters (including alkyds), the acrylics and vinyls, the epoxies and polyurethanes, phenolics and aminos. End-user classifications are often adopted for market analysis. Uses are commonly divided into three categories: architectural paints (including decorative and building paints), the production line finishes, (or OEM - original equipment manufacture - finishes) and a myriad of special types. The following discussion will respect this latter classification.

Architectural paints are the largest segment of the market, accounting for over 55% of paint sales in the EU market. Architectural and decorative paints have reached a stage of technical maturity after a number of dramatic transformations, starting in the 1950s and 1960s, spurred by the introduction of new components: the arrival of alkyds and titanium dioxide and later of vinyl and acrylic emulsions steadily improved product performance in terms of dripping, drying, application and durability (Coatings Comet, 2001). New manufacturing methods and retailing techniques have also greatly reduced costs of production. Hence, over the years, the industry became a rather mature segment dominated by price competition (exacerbated in more recent years by the rise of cut-price DIY retailers) and shrinking profit margins. Achieving high volumes in production and distribution is the key to survival. Aggressive marketing campaigns are also used to differentiate products. The decorative paints business is subdivided into trade and retail paints. The main customers for trade decorative paints are professional painters and decorators and housing contractors. As these customers vary from relatively large companies to self-employed individual painters, the market is serviced in a variety of ways, from direct sales for the bigger customers to reliance on large retail chains for the rest. The main players in this segment are ICI, Akzo Nobel and Kalon. Some SMEs and industrial coatings manufacturers retain a stake in this market. Retail decorative paint customers are mainly individual DIY enthusiasts, who buy the product in small-specialized shops or from large retail chains. National specificities matter a lot in this respect. The main players in this segment are Sherwin Williams, and again ICI and Akzo Nobel. Akzo Nobel has reached a position of global leadership in terms of sales via a rather intensive program of mergers and acquisitions. Akzo's still recent acquisition of Courtaulds assets in the business caused some upheaval in the top firms rankings (see Table 1). ICI Paints, however, has a more global sales pattern, with Akzo still relatively specialized in the Western and increasingly Eastern European market. The merger between Kalon and Euridep created another major player, whose market share has been achieved mainly by producing own-label paints for the big DIY retailers

In looking at industrial applications, product performance (life-time, durability, thinner coatings, UV resistance, environmental compliance etc.) and application processes matter as much as (or even more than) price. In the OEM segment, producers rely on long-term agreements with their customers, mainly automotive companies, who have proved to be very demanding customers, spurring a process of continuous innovation to increase both performance, range of products, and environmental performance. More generally, most users of industrial coatings have aimed to achieve better coordination of painting and dying with other manufacturing steps in order to reduce manufacturing costs and decrease process downtime. Henry Ford's memorable motto ('You can have any color you want, as long as it's black') highlighted one key bottleneck in the automotive industry. From the mid 1980s, US companies took a clear lead in developing long-term deals with their customers.

Finally, more sophisticated, high performance coatings are segmented in a variety of end user markets, from aerospace to underwater marine applications. This is by far the most science intensive part of the industry, with high R&D costs, high value added, and high uncertainty. Developing paints for this segment requires systematic investments in R&D, dedicated production facilities, and risk taking attitude. Demand is usually limited. However, the potential advantages in terms of R&D spillovers and reputation effects are extremely high, although hardly quantifiable. For example, Courtaulds Coatings used to be the leader for high performing marine coatings. Executives of the company attributed this lead to relationships with their aerospace division. However, in the aftermath of Akzo Nobel's acquisition of Courtaulds Coatings the aerospace unit had to be disposed of to avoid anti-trust litigation at the EU level.

3. INDUSTRY STRUCTURE IN THE 1990S

Within the context of the EU manufacturing sector, the paints, coatings and printing inks industry represents a sizeable share of the overall chemicals industry. Throughout the 1990s, this segment accounted for about 6% of the chemical industry's total turnover, with a similar share in terms of total R&D spending. As a matter of comparison, in 1998 pharmaceuticals and basic chemicals accounted for 38% and 31% respectively (Panorama, various years;

CEFIC - European Chemical Industry Council, ditto). The EU countries have traditionally been the world's largest producers, followed by the US and Japan. However, the Asia-Pacific region is now the fastest growing area in terms of production. Throughout the 1990s, the paints and coatings industry underwent a long process of restructuring and consolidation. In the first half of the 1990s, the efforts of large chemical firms to shift their scientific and technological focus from traditional chemistry to life sciences sparked a series of (de)Mergers and Acquisitions (M&As) that reshaped the corporate landscape of the industry. The examples of ICI and Zeneca in the UK, Hoechst and Herberts (later acquired by DuPont Performance Coatings) in Germany, Courtaulds acquired by Akzo Nobel in the UK and Netherlands exemplify this trend. Nowadays, the industry (and particularly the decorative paints end of it) is dominated by a relatively small number of very large firms that compete on a global basis within their particular market segment.

This industry tends to mirror GDP fluctuations. In particular, the sector is anchored to the cycle of its main industrial users, namely carmakers and the construction industry. Consequently, the last three years have been quite tough for the whole sector, after a decade of overall positive results. From the beginning of 2001, the financial results of most companies started deteriorating, and all of the major players in both the EU and US (Akzo Nobel, BASF, ICI, Bayer, Dow, DuPont, Rohm and Haas and Sherwin Williams) declared quarter-by-quarter worsening financial performance (Coatings Comet, 2001). Since then, most EU producers have announced rationalization programs aimed at reducing overcapacity. This has generated downward pressure on both employment and capital expenditure. Moreover, the wave of M&As slowed considerably toward the end of the 1990s. Most of the recent deals in Europe and North America concern second-tier (in terms of size) companies, usually within national boundaries, and in the specialty coatings sector. Developments in China have displayed a different trend, largely due to China's recent entry into the WTO. This prompted local firms to reorganize to be ready to face the challenge of foreign producers that now can enter through fully owned holdings, replacing the previously favored joint ventures approach (Coatings Comet, 2001). Table 1 reports the top producers by sales in various years. It is estimated that the top twelve firms listed represent about 40% of the entire industry (by sales).

In terms of strategy, most companies have decisively adopted a 'core business' approach. Within this core they struggle for global leadership. The recent slow down has enhanced this trend that was already quite visible, particularly in the US, around the mid 1990s. So, for example, in 1993 ICI Paints sold its powder coating business to Ferro, its automotive OEM unit to

DuPont in 1994 and its vehicle refinishing business to PPG. Nowadays, ICI maintains a leadership position in decorative and packaging coatings only. US companies shopped around in Europe throughout the 1990s, mainly within their core business areas though, and benefiting in particular from the increased pace of restructuring in the German chemicals industry in the mid 1990s. During this period PPG acquired ICI assets reinforcing their leading position in the automotive business. DuPont acquired Herberts from Hoechst in March 1999. PPG also expanded its industrial coatings concern by acquiring BASF's packaging coatings business. Generally speaking, European firms have lagged behind their US competitors in these efforts to specialize in specific market segments. In the aftermath of the last recent wave of acquisitions, both observers and practitioners expect a great deal of rationalization to take place within, rather than between companies. As said above, the most recent news from the industry focuses on the reorganization of internal company structure, the revamping of old plants, and closure of old units. In terms of growth, most expansion projects are located in the still growing markets of the Asia-Pacific regions, and especially China.

*Table 1. Top twelve companies by sales. All data in billion US$. * estimated.*

Company	*1999*	*2001*	*2003*
Akzo Nobel (NL)	5.1	5.15	5.33
PPG Industries (USA)	3.4	4.66	4.8
ICI Group (UK)	3.5	4.36	4.39*
Sherwin-Williams (USA)	4.9	4.31	4.25
Henkel (G)	2.76	2.78	3.85
DuPont Performance Coatings (USA)	2.4	3.49	3.6*
Valspar (USA)	1.15	1.48	2.12
BASF Coatings (G)	2.3	2.07	1.98
RPM, Inc. (USA)	1.61	1.95	1.98
SigmaKalon (NL)	0.89	1.54	1.6
Nippon Paint (J)	1.6	1.58	1.52
Kansai Paint (J)	1.5	1.46	1.47

Source: http://www. coatingsworld.com/; last accessed July 31 2003.

In terms of industry structure, it is also important to bear in mind that traditionally a huge number of small and medium firms co-exist with the global players. They survive by specializing in niche markets that are large enough to generate economies of scale, but demand rather specialized competencies. Examples of these niches are solar reflective paints, or powder coatings. For example, medium sized companies have established themselves as world leaders in niche markets that demand specific R&D efforts, such as Ferro in powder coatings. However, there are high barriers to entry that limit the opportunities for small, specialized new entrants to challenge the leadership position of large incumbents in high volume product markets. The barriers do not seem to be the development of the relevant scientific and technological capabilities. First, there are problems in the application of these new kinds of paints that inhibit their introduction in to large volume markets. Indeed, most medium sized specialist companies sell intermediary products to industrial users. Second, accessing the distribution network, or establishing an entirely new one, is extremely difficult. Large firms rely on their own outlets, but also on large retail centers (e.g. B&Q in the UK). These large retailers are unlikely to offer shelf space to products untested on the market and, in addition, are increasingly marketing own brand products, usually relying on medium, but already established manufacturers. Similarly, in the industrial coatings sector, car manufacturers demand global partnership agreements and an increasing range of services. New entrants are unlikely to be able to provide these capabilities, either in terms of scale of manufacturing or scope of services required.

4. INNOVATION: SOURCES, TRENDS, AND APPROPRIABILITY

As far as the chemistry underlying the final products is concerned, radical changes in formulations seldom occur - and rarely in the decorative and industrial coatings business. Changes tend to be incremental with the aim of achieving a better combination of the usual components (resins, pigments, solvents, and additives). Production of specialty coatings, however, displays dynamics closer to those of other 'science intensive' industries, and thus is led to a much greater extent by scientific and technological developments. This is not to say that innovation in the decorative coatings business is easy. The difficulties of any innovative effort are inherent. Rather, the sources of these difficulties are different: market uncertainty matters to a much greater extent than technological uncertainty. Accordingly, the roles of marketing and market research are much more important. Even in those sectors, such as vehicle paints, where the customers are firms (i.e. carmakers), a clear understanding of the preferences of end customers is fundamental. Besides, and with the exception of the specialties end of the industry, processes are fairly well understood. The process steps necessary to combine the raw materials involved in paints production entails a lower degree of technical difficulty than is the case for instance for pharmaceuticals or petrochemicals. Pressures and temperatures involved are much lower, reactions are often no more than stirring operations that may not even require catalysts. Quite commonly, the chemical and process engineering capabilities required have more to do with scheduling and plant management than design. Accordingly, innovative contributions usually occur up front, as they stem from the interaction between R&D activities and marketing. Indeed, it is the role of marketing that explains why the Community Innovation Surveys (CIS) highlighted the disproportionately high number of product innovations that characterize this relatively mature industry (WZB, 1996).

4.1 The drivers

Innovative efforts in this industry, including the fairly mature decorative paints market, were rekindled in the 1990s by the interplay between two key factors. First, environmental regulation has played a pivotal role in driving industry R&D (and lobbying) strategies. In particular, throughout the 1990s, VOC emissions regulation became of paramount importance. VOCs are compounds emitted by the solvents during drying out. The efforts of EU and USA regulatory authorities to cut these emissions have opened up new avenues for research. High solid coatings, powder coatings, waterbased, UV/EB (ultra-violet/electron beamed) radiation cured coatings are the most promising new research routes. Powder coatings are often seen as the most promising for investigation. So far, these new technologies are making their impact mainly in non-architectural coatings. A few companies have achieved leading positions in these systems. For instance, in the late 1990s five companies (Akzo, BASF, Courtaulds – now Akzo; and Herberts – now DuPont, together with the US Ferro) controlled over 50% of the powder coatings market in Europe. Ferro is also a leading actor in the US market. Similarly, in the radiation-curing sector the scene is dominated by the producers of monomers and oligomers, such as the Belgian group UCB, rather than by the large paint manufacturers. Companies such as ICI, DuPont and Sherwin Williams are leaders in waterbased and high solids.

The second key driver of change has to do with the role of customers in driving producers' search for new raw materials, which combine performance in application, reliability in production (key issue in the OEM market) and contained costs. For example, in the OEM business, car manufacturers took a strong stance toward the development of more efficient (in terms of cost), effective (in terms of performance and variety) and environmentally friendly (particularly with respect to VOC emissions) techniques. Time efficient painting/drying systems have been designed to satisfy the carmakers' quest for flexibility. Further, paint manufacturers have been offering bigger color ranges and better paint durability. US manufacturers of OEM have been particularly aggressive in seeking exclusive agreements with car manufacturers to establish long-term relationships. On the one side, the car manufacturers are rewarded with efforts dedicated to design of a paint system tailored to their own production facilities. On the other side, the paint manufacturer locks the carmaker into its own process. In the decorative paints business too quality is increasingly becoming an issue. The first upsurge of VOC-free compounds has often been accompanied by a decline in performance (e.g. thickness and resistance of the coating film) and extra-costs to the manufacturers. There are technical problems that still need to be resolved, particularly in terms of production processes, to make emerging technologies as reliable and efficient as the older ones. For instance, the successful introduction of thin film powder coatings into the domestic appliance market is vital if powder coatings are to fulfill their promise. Yet, developing a production process capable of guaranteeing the film thickness necessary to compete with established processes (e.g. coil coatings) has still to be achieved.

Overall, environmental regulation and ever more demanding customers have forced the industry to change mentality. Companies are abandoning their traditional product-oriented focus to adopt a process- (or system-) oriented approach. In fact, the availability of new raw materials (whether pigments, solvents, additives or resins), tougher regulatory frameworks, and shrinking margins, have increased the number of factors it is necessary to consider when selecting a coating and its production process. Traditionally, choices were based upon a combination of cost and expected performance. Nowadays, manufacturers have to take account of regulations related to the environment, health and safety, fast changing customer needs, new emerging markets, and increasing complexity of the knowledge base. However, as will be discussed below (section 7), different segments of this industry have reacted differently to the interplay of these two key drivers of change.

4.2 Appropriability and innovation

Imitative processes in the paint industry are relatively simpler to achieve than in other sectors. However, the reverse engineering of paints still is an activity that requires highly specialized skills (Interviews, 1999). It is not easy to disentangle the specific characteristics of each component (of the resins in particular). Nor is it easy to acquire the level of skills necessary to effectively mix the resin adopted with the other components. Effective mixing is fundamental for both color delivery (and thus the quality of the final product), and cost efficiency: mixing mistakes can often be emended by increasing the amount of pigment used, but as pigments are the most expensive raw material, the impact of costs can be punitive. The most complex and surely the least transparent stage of chemicals manufacture is the reaction stage, in which raw materials and catalysts are mixed to obtain the final product. However, in paint production the reactions involved are in the main 'mixing and stirring' operations, within manageable pressure and temperature ranges. Suppliers and producers of raw materials are more likely to come up with patentable innovations. Indeed, the dyes and pigments sector is exhibiting quite lively innovative dynamics (Interview, 1999).

Technological difficulties aside, appropriability can be achieved by alternative means. One of the chief channels is through the control of the distribution network. This is the case in The Netherlands, for example (Kemp, 1993), where producers generally have their own distribution networks. In the UK and Germany, architectural paints are usually marketed through the wholesale trade. However, the largest manufactures have traditionally maintained their own outlets. In addition, large paint manufacturers are pushing their way into these retail networks, and commercializing flexible 'color matching' equipment (tinting systems), operated by the retailer personnel. Consumers can make up the colors they prefer (from an extensive range of combinations). The color is then obtained by combining a white base with the pigments provided by the paint manufacturers. Consumers' choice is increased, retailers offer an additional service, and producers lock the distributors into their products. This is not a radical innovation in distribution techniques, though: tinting systems have been around for at least a couple of decades (Interview, 1999). However, color ranges and specifications have surely grown in size and quality. Further, the equipment has greatly improved, as the delivery of the color chosen from a catalogue is much less dependent on the skills of the operator. Kemp (1993) also points out that direct control of the distribution network can have negative effects in terms of 'product cannibalization'. Indeed, innovating firms that sell their products through their own channels reach the same pool of clients. Thus, rather than increasing their market penetration, they may end up crowding out their existing products.

5. ENVIRONMENTAL REGULATION AND INNOVATION

The issue of what impact environmental regulation may have on firms' innovative behavior has been extensively analyzed. Rest and Ashford (1988), in their review of the literature, discuss a wide range of issues that touch upon the relationship between innovation and regulation. In particular, they discuss the evidence that a positive correlation can be established between regulation inspired by environmental concerns and innovation. First, as regulations set strict standards for product performance, commercial risk may actually be reduced by intensifying pre-market testing requirements. Second, regulation can give a clear focus to otherwise undirected innovative efforts, for instance by making firms aware of the need to develop compliant technologies. This could reduce the uncertainties associated to R&D projects. In this connection, it has been observed that regulation-induced innovations have led to improvements that go far beyond the specific objectives of the legislation (Ashford, Heaton and Priest, 1979). This probably occurs because regulations provide incentives for mature industries to innovate, forcing them to look at established products and processes, acquire relevant capabilities and thus increase the likelihood of innovation, whether compliance related or not. Finally, regulation can create market opportunities for small, innovative firms able to introduce safer products. Firms capable of anticipating regulation will enjoy first mover advantage or, even better, might set the industry standard if their technology is identified as the 'best available technology' by the regulator (Skea, 1995).

Along these lines, a number of studies have stressed and even quantified the positive effects that regulation can have on innovative behavior. A well-known example is the role played by government intervention in making firms aware of the advantages of picking the 'low hanging fruits'. Schmidheiny (1992) studied the case of 3M's Pollution Prevention Pays program. While achieving substantial reduction in solid waste, wastewater and emissions to the atmosphere, this program also delivered total savings of some $530 million. Other examples can be found in the chemical industry in general (Dorfman *et al.*, 1992) and in the titanium dioxide industry in particular (Groenewegen and Vergragt, 1991). Despite this, it continues to be very difficult to identify general conditions capable of determining 'optimal' regulatory strategy. Rest and Ashford (1988) point out that the end results are likely to vary according to the specific characteristics and circumstances of the sector that is the object of regulation, as well as to the specific objectives that inspired the government action. They put forward a framework aimed at identifying the effects of regulation according to the life cycle stage reached by the productive segment that is the object of regulation.

In particular, they rely on the framework suggested by Abernathy and Utterback (1975, 1978). This recognized model predicts an early stage of fluid and continuous product level innovations that lead to the establishment of a dominant design. This, in turn, leads to an industry shake out, a reduction in the number of firms, a dramatic decline in the initially high rate of entry and exit (which was very high in the early stage of the life cycle). This is accompanied by a decreasing rate of innovation in general, and in particular by an increasing relevance of process as opposed to product innovation. The fundamental point to be stressed here is that the regulatory action does not need to be tailored to the set of technological capabilities *currently* available to the industry. Rather, the regulator should analyze the sector innovative *dynamics*. This would allow for a proper identification of the most suitable form of regulation. In line with this, Ashford, Ayers and Stone (1985: 422) stress that a sensible assessment should rely on:

> ...a historical examination of the pattern of innovation in the regulated industry, an evaluation of the technological capabilities of related sectors having incentives to develop compliance or substitute technology, and a comparison between the regulated sector and analogous sectors with documented technological responses to regulation.

As the sources of innovation and the nature of innovation itself change along the product life cycle, the regulator needs to take into account the 'biography' of the product, and tailor its intervention accordingly. Further, different degrees of stringency may alter the outcome of the regulations. Indeed, most studies pinpoint *stringency* as the one most important criterion to assess the legislator's capability to activate innovative search in firms. A regulation is stringent inasmuch as it demands a significant reduction in emissions or decreased use of substances identified as hazardous or toxic. In turn, this reduction is achievable by introducing a significantly new product or process, as compliance using the established technology is generally costly. The empirical evidence surveyed by Ashton *et al.* (1985) confirms that product regulations tend to determine product innovations, while process regulations are usually met by process innovations. Most of all, stringency comes out as the most important factor in determining whether or not regulations call for innovative answers. In the sections below, the relevance of stringency will be further discussed and qualified.

6. THE DEVELOPMENT OF LOW SOLVENTS AND SOLVENT-FREE COATINGS

This section will discuss the evolution of the legislation on VOCs and the existing methods to cut VOC emissions. It will analyze the specific issue of the diffusion of waterbased paints in two segments of the industry, namely decorative paints and vehicle paints. In the UK, the Department of Trade and Industry considers that the paint and print industries constitute about 10% of total emissions. Petrol engines are the biggest contributors to emissions (about a quarter of the total), closely followed by the oil industry and agriculture (Table 2 shows the relative rankings).

Table 2. Contribution to VOCs emissions by manufacturing sector in UK.

Source	Emission kt/a
Solvent user industries: total Of which:	734.7
Paint industries	261.3
Print industries	41.3
Metal industries	42.6
Dry cleaning	11.0
Adhesives	58.0
Pharmaceuticals	40.0
Aerosol consumer products	85.6
Non-aerosol consumer products	73.0
Agrochemicals	39.0
Seed oil extraction	10.0
Other solvent users	70.6
Oil industry	532.0
Chemical industry	200.0
Stationary combustion	86.4
Food industry	74.7
Iron and steel industry	9.1
Waste disposal	29.0
Agriculture	297.5
Miscellaneous stationary sources	39.9
Mobile sources	972.0
Total all sources	2974.6

Source: DTI internal report, 1999.

VOC emitting operations were probably the first industrial activity to attract the eye of the regulator. In the UK, the Alkali and Works Regulation Act dates back to 1906. In 1999, the EU issued a specific directive regulating VOC emissions from manufacturing concerns. The interviewees, the source literature reviewed and the CEPE (European Council of the Paint, Printing Ink and Artists Colours Industry) do not consider this directive as particularly 'threatening' to the industry. Interestingly, this position represents a major change in industry's attitude toward environmental regulation. As late as the mid 1990s, this kind of legislation was most likely to be considered as a threat to manufacturers' competitiveness (Interview, 1999). However, in more recent years, the industry has adopted a more proactive role, and has tried to influence and shape the regulatory process, rather than passively await its outcome (Giordan, 1995). In the decorative paints business, as will be elaborated below, regulation has become to be considered as an integral part of the innovation process, rather than as an external shock (Interview, 1999).

The recent EU directive can hardly be seen as a threat to competitiveness. Indeed, it openly relies on the concept of the 'best available technology'. It sets a performance standard that must be achieved before the end of October 2007 (European Commission, 1999). However, the directive does not set any specification standards. That is to say, it does not explicitly push companies to adopt waterbased formulations: as long as the emissions target can be met by a reduction in the amounts of solvents in traditional products, firms are free to continue to use them. Finally, implementation of the regulation is left to national governments. Thus, the actual stringency of the directive is likely to depend on the outcome of the 'bargaining' process between the industry (via its associations) and the government involved. In this respect, the most contentious issue is related to the techniques to be applied to measure the level of emissions (Interview, 1999). This process is likely to generate discrepancies in implementation at national level. It is worth noting here that the existing regulation on VOC emissions impacts on the production process for paints and coatings. Presently, there is no regulation (at European level) related to the final product. Country level legislation has also tended to neglect this aspect.

Different techniques are available to cut VOC emissions. End-of-pipe technologies of course were the first solution introduced by the industry, and still are the most common (Interview, 1999). Various filtration, destruction and recycling techniques continue to be applied in decorative paints manufacturing plants. Akehurst (1997) discussed in detail the various techniques available to destroy VOCs. First, he considered thermal oxidation and catalytic incineration.

Both techniques involve combusting the hydrocarbon molecules that produce the emissions in order to convert them into carbon dioxide and water. The former technique requires high temperatures and destruction of the VOCs takes a relatively long time. This method is used in those sectors where heat is useful and can be 'shared' with other manufacturing processes. It is quite commonly used in the petrochemicals, plastics, computer and motor industries. The paints and print industries rely more on catalytic incineration where combustion occurs more quickly and at lower temperatures. However, this technique is more capital intensive than thermal oxidation. Separation alternatives are also available. In the paints and printing inks industry, adsorption of gaseous organic material via activated carbon is a very common procedure for solvent recovery. The main problem here is disposing of the VOCs recovered. Recycling into the manufacturing process itself is possible, but only up to a certain point. Quite often they are passed to specialized firms for disposal. Biofilters, in which pollutants are 'digested' by bacteria and fungi, which separate them into carbon dioxide and water are gradually becoming more common in the industry, but their novelty makes adoption of this technique a rather 'adventurous' endeavor. . The result of this technique is an accumulation of dead organisms that eventually must be removed and disposed of. The following section focuses on the specific case of waterborne paints, as an instance of 'clean' technology, as it removes from the product the main source of VOC emissions.

7. THE ADOPTION OF WATERBORNE FORMULATIONS

The discussion of the two cases presented below (diffusion of waterbased paints in the decorative and industrial paints businesses) should be seen in the context of the model discussed by Ashford, Ayers and Stone (1985). Paints are a fairly mature line of business, which is nevertheless characterized by the persistency of product innovations. It is argued that innovations in the decorative paints business might have more to do with marketing and creation of new customer niches, than with new formulations. However, new formulations can be achieved, and raw materials' suppliers exist to provide inputs to producers' innovative efforts. What matters most to this industry though, is customers' needs and access to customers (Interview, 1999). In addition to supply side considerations, the demand for cleaner products and technologies is pivotal in this industry. Accordingly, the examples presented below focus on the role played by customers in shaping the industry response to the emergence of cleaner technologies (in terms of VOCs). It will be noted that understanding demand characteristics and their evolution is a vital factor in successful innovation in this industry. Further, and relatedly, the role of 'stringent' regulations will be stressed and qualified. It will be seen that the main role of stringent regulation should be to influence demand patterns. Changing consumers' behavior and attitudes seems to be the best way to influence firms' behavior.

7.1 The vehicle coatings sector: proactive customers

The vehicle coatings sector has showed a very dynamic response to the emergence of waterbased technologies mainly led by the efforts of carmakers to respond to environmental pressures. The main European players are PPG, BASF, Herberts, Akzo and IDAC. PPG is the world leader, thanks to its strong grip on the US market. In 1996, PPG alone supplied nearly half of the entire US market. This gives PPG the additional advantage that they can supply US producers world wide. Traditionally, the division of labor between car manufactures and paint suppliers was quite clearcut. Once ready for treatment, car bodies were shipped to the paint supplier in batches. The cars were returned to the manufacturers' production units for refinishing and inspection. The diffusion of just-in-time techniques throughout the automobile industry supply chain has not left the paint process untouched. However, paint suppliers were affected much later than most other component suppliers and their process of adjustment did not take on any momentum before the early 1990s.

Under pressure from Japanese competitors, in the 1980s US manufacturers reacted by adopting, and adapting, their organizational practices and production techniques. European producers did not react so immediately. Paint shops were the main casualties of these changes. Two problems had to be faced. First, the traditional system was based on huge inventories capable of buffering the lack of co-ordination between manufacturing steps. Second, the range and quality of colors had to be greatly increased to meet raising customers' expectations. The first problem required a change in the working relationship between paint suppliers and carmakers. The second required a good deal of technical change, usually pursued on an alliance or joint venture base. Given the concentration of the purchasing power, paint producers could do nothing but adapt to the new rules of the game. In the 1990s the automobile industry has deliberately approached this issue from an environmental point of view. The industry itself was the subject of increasing environmental concern and cost considerations (for labor as well as material) pushed them to adopt more efficient technologies in their paint shops.

In order to increase the quality and reliability of the process, car producers have also shown a tendency toward decreasing the number of their suppliers and giving more and more responsibilities to the few that they use. Increasingly, car makers expect paint manufacturers to take on the responsibility of global suppliers for all of their manufacturing plants. This represents a major break with the past, when car manufacturers were used to rely on a large number of suppliers. Although globalization entails some problems for suppliers, it guarantee them a degree of appropriability of their innovative efforts, as fine tuning the plant to the specific characteristics of the paint provided can take several years. A carmaker's choice of a particular paint manufacturer is a major event. Paint shops are one of the most polluting stages in automobile production and therefore are a very important consideration in the design of a new automobile manufacturing plant. The choice of paint supplier is a strategic choice involving long-term commitment.

Historically, this process has always represented a major bottleneck in the production chain for carmakers. The painting process accounts for about 10% of the total manufacturing cost of a car and, in terms of logistics and plant management, the process is a disruption to the throughput. Traditionally, this has been handled by sending car bodies to the paint supplier's plant. Henry Ford's famous remark, rather than being seen as an example of manufacturer sovereignty, should rather be seen to be an acknowledgement of the difficulties involved in the painting process.

The process involves a standard metallic paint system being applied to the car body in four different coatings. First, a layer of primer is applied by electro-coating the car body to protect the main body from corrosion. Second, a surfacer coat (the first layer of paint) is applied to prevent stone chipping and to smooth uneven surfaces. Then the topcoat that gives the final color is applied and finally a clear coat (i.e. a varnish) is applied to produce the gloss effect and guard against weathering. This topcoat was the chosen candidate for becoming a water-based paint. Indeed, the topcoat uses the largest quantity of paint, and thus solvents. The adoption of waterbased formulations would therefore imply a substantial reduction in VOC emissions. However, solvents must still be used to clean pipes between color changes. Further, despite the progress that has been made in reduction of VOC emissions, numerous hazardous substances are involved in the electro-coat, surfacer and clear coats. The clear coat is probably the worst offender, as it involves arsenium K2, a very dangerous product. The emphasis nowadays given to wide ranges of colors, metallic and glossy effects, and durability may well offset the overall advantages of the adoption of waterborne topcoats, because, in order to reach the required thickness, base and clearcoats have to be extensively used. Electro coat application remains a necessary and energy intensive stage. Widening the colour range necessarily means that pipes need to be cleaned more often, which, in turn, involves the use of solvent-based cleaners. In theory, in order to achieve the biggest reduction in VOC emissions, the best solution would be a combination of a cathodic electro-coat, a powder primer, a waterbased topcoat and a powder clear coat. In practice, powder paint technology still needs further development to provide the quality required by the market.

7.2 The decorative coatings sector.

The previous paragraph highlighted how a rather complex application process was adapted to face environmental and cost pressures hit the main users of paints manufacturers. The response was relatively fast, and environmentally friendly. This and the next sections focus instead on a relatively simpler product and process, which took longer t react. Organic solvents were first adopted in the 1950s, when synthetic binding-based products begun to offer a technically viable alternative to traditional linseed oil-based paints. A basic ingredient of the new formulations was some variant of a liquid organic solvent, used to increase the spreading properties of the paint and released during its production and application. These emissions have negative effects on the environment (e.g. they contribute to the formation of smog) as well as on the users' health (e.g. in the short term they worsen the symptoms of asthma, while prolonged exposure has carcinogenic effects). Thus, the main paint manufacturers started developing alternative, low solvent products early on.

Nowadays, both solvent based and waterbased co-exist in different niche markets. In terms of manufacturing processes and formulation, the key difference is related to the making of the resin. Alkyd resins are employed to produce gloss (solvent based) paints and deliver particular properties such as non-drip effects, and water resistance. They are produced by mixing various substances at atmospheric pressure and relatively high temperatures, ranging from 240° C during the early reaction steps, down to 50° C when the resulting alkyd is stored. This kind of endothermic reaction involves little risk of getting out of control. Once the alkyd resin is ready, it is mixed with the pigment (titanium dioxide for whites, and other organic pigments for colors), solvents (such as white spirit) and additives (the most common being driers and anti-skin solvents). The most expensive component is the pigment. The most sensitive operation at this point is the mixing of the various components in such a way as to reach optimal dispersion of the pigment particles. Although technically difficult, this operation is not dangerous, being only a series of stirring phases.

Making a resin for waterbased paint is a rather different process. The resin is actually formed as an 'emulsion' of oil and water and brings the main advantage of requiring water as solvent. About 50% of the resin is water. Mixing a surfactant (i.e. soap) with water brings about the basic reaction, at 70° C. Emulsion paint dries because the water it contains evaporates. In order to improve the film strength, a high boiling point solvent is added. Biocides have to be used to stop the growth of bacteria. Further, other additives have to be used to guarantee ease of application in indoor, non-professional applications. The most common additives adopted aim at aiding dispersion, thickening the film, avoiding foam formation, favoring drying at low temperatures, and so forth. Extenders have also to be used in order to reduce the amount of pigment necessary (still the most expensive component). As a result, the formulation of emulsion paints tends to be more complex than that of alkyd paint.

The paint industry has always been seen as rather conservative. Despite the continuous experimentation with new formulations, only a few reach the market (Kemp, 1993). The main problem is associated with the lack of demand for 'cleaner products' (Interviews, 1999). While the development of fast drying (high solvent) paints was pulled by the automobile industry, the development of low solvents paints has mainly been the result of regulatory pressures (and, sometimes, the sheer threat of regulations) rather then the product itself. Indeed, the architectural market has been quite hard to penetrate, given the well established preferences of do-it-yourself users. Gloss, durability and ease of application have always driven this market segment. Low solvent paints (such as waterbased paints) have not always matched up with these requirements.

7.3 Sources and organization of innovation in the decorative business

Innovations in the decorative business do not necessarily entail new formulations or changing chemistries. They may well stem from marketing and commercial efforts, the adoption new information systems, and the introduction of new accessories (e.g. rollers). Indeed, product innovations appear often as commercial innovations, e.g. new color range such as natural whites. One of the key objectives of 'market led' innovations is to define a set of consumers served by 'standard' products as an independent, more specific market and occupy it before rivals (Interviews, 1999).

The drivers for developing new formulations are related mainly to environmental considerations (e.g. low solvent paints) and customer tastes (e.g. more or less gloss). As both are strongly related to local market characteristics, the division of R&D labor strongly resembles that discussed by von Hippel (1995), whereby the need to access 'sticky information' leads firms to localize their R&D and/or production facilities in the proximity of the final customers. Furthermore, a new formulation should be seen as a new balance among the usual components, rather than a radically new chemistry. In the case studied, the company HQ plans to launch at least two-three new products a year (Interview, 1999). The initiative is usually taken by the market research department - the so-called 'innovation groups'. These latter involve people from different backgrounds with different responsibilities in the company. However, the key insights are likely to come from market research and R&D, which has been identified as the key interface for innovative activities (Interviews, 1999).

Suppliers are occasionally involved in developing a new formulation. This usually takes place via contracts and secrecy agreements. Other partnering forms such as joint ventures and alliances do not usually occur, as most innovations are incremental in technological terms (although some trade journals report an increasing reliance on R&D joint ventures). Also, innovations stemming from raw materials other than resins are unlikely to generate major changes in the chemistry (Interview, 1999). Resins are the component that determines the performance and cost of the final product, and their production is most likely to be kept in house by large manufacturers. In short, incremental improvements can be dealt with by exchanging material specifications under secrecy agreements with suppliers.

As for the organization of innovative efforts, paints producers have to respond to remarkably strong national specificities, in terms of legislation related to manufacturing operations and customer preferences. However, globalization does not have determined product standardization. Thus, large manufacturers in general, and the one analyzed here in particular, do rely on centralized R&D departments (formulations are homogeneous), but they still need decentralized production facilities to tailor the final product to local tastes (Interviews, 1999). The fact that production facilities have to be located close by the final market, reduces the competitive implications of heterogeneous environmental legislation. As long as environmental regulation impinges upon the manufacturing process, firms that relocate their production facilities in the proximity of the final market compete on a fair basis (at least in terms of the regulations they face). Interviewees confirmed that most of the existing process-related legislation could actually be met by improving the process itself, leaving the product formulation unchanged.

In this respect, the great divide is between the US and Europe. Indeed, all the interviewees pointed out decorative paints are used rather differently in the US and the EU. To quote one interviewee:

> 'In the USA there is more wooden paneling on the outsides of buildings, and for such large surface areas, water-based gloss paints are more appropriate. In Europe there is less wood on the outsides of buildings, but there is more of a need for high sheen level, hence the preference with consumers for solvent-based gloss paints' (Interview, 1999).

Tastes vary as well: gloss effects are not very popular in the US, thus the formulations, and necessarily resins, used are different. The main result of these differences is that, traditionally, waterbased products are more diffused in the US thanks to 'natural' market demand, rather than regulation. Therefore, US companies are benefiting from the advantages stemming from early adoption of this particular kind of formulation. The adoption of waterbased formulations in the EU countries is being slowed by demand conditions, rather than lack of technologic capabilities (Interview, 1999).

In fact, the incentives to reduce VOC emissions seem not to be regulatory led, bur result from the efforts of the industry to avoid 'bad publicity' and maintain competitiveness (Interview, 1999). The real problem is that consumers are conservative and waterbased paints are penetrating slowly. It is interesting to note that the industry (via its European association, CEPE) is lobbying to get product level regulations introduced at EU level. Individual companies are also active at national level launching marketing campaign to make customers aware of the issue of VOC emissions. Regulations should aim at changing the pattern of demand, so that all paint producers change at the same time: 'Otherwise it would be commercial suicide' (Interview, 1999).

It is worth noting that there are alternative to water based formulations which provide solutions to some of the problems that are inherent in the use of these paints, for example, that the water evaporates very quickly and favors the development of micro organisms. However, these alternatives increase the technical difficulties. VOC emissions can be reduced by changing the formulation of the resin (for instance, by reducing the content of white spirit), but this increases the final cost of the product, partly because the resin becomes more expensive per se, and partly because other additives are needed to achieve the same performance of standard products (particularly in terms of spreadability and durability). The trade off here is quite steep. Overall, it is fair to conclude that, whatever route is chosen, the science content of paints is increasing (Interview, 1999).

7.4 The impact of regulation

Section 6 discussed various techniques to reduce VOC emissions. A number of end-of-pipe solutions are available to producers. It was also said that the adoption of waterbased formulations (i.e. an example of 'cleaner technology' as it minimizes the actual use of solvents) would represent a dramatic step forward in terms of VOC emission abatement. From the interviews it is clear that there is no natural progression from the adoption of end-of-pipe solutions to the introduction of cleaner technologies in the case studied. In short, it seems that the incentives and the capabilities required to implement specific solutions are quite different.

First of all, EU regulation relates only to emissions caused by the manufacturing process. Product emission reductions (i.e. the product's content of solvent that will evaporate during manufacturing and application) are done on a voluntary basis. Therefore, the diffusion of waterbased paints has been delayed more by the characteristics of the consumer base, than by the lack of technical capabilities. At the process level, VOC emissions reductions are the outcome of intra firm efforts that the customer is totally unaware of. These efforts have surely been positively influenced by the putting in place of regulations or by just the threat of them. It is fair to say that the industry has largely anticipated actual government intervention. However, process regulation has brought about process innovation only (as would be expected). Further, this kind of improvement is entirely intra-firm. Consumers do not perceive that anything has been changed. This has the great advantage of decoupling the adoption of VOC reducing techniques and the marketing strategies of manufacturers (a very sensitive issue). On the other hand, there are limits to what can actually be achieved in terms of VOC emission abatement.

As said above, the industry itself is lobbying at European level for a shift in demand patterns that will make the adoption of waterbased formulations economically viable and less risky. In the US, some states (e.g. California) have already put in place some kind of product level regulations, limiting the VOC content in the paints sold. This is incidentally relatively easy to do in areas such as California because the climate does not vary much across the state (and climate is a key variable in defining the characteristics of the resin to be adopted). More recently federal measures have also been introduced (the upper limit for VOCs varies both according to product and to the feasibility of its being achieved). As federal regulations have to accommodate for a wide range of climates from the extreme cold of Alaska to the temperate climate of Florida, the limits they set are more relaxed than those adopted by state laws (Interview, 1999).

8. DISCUSSION AND CONCLUSIONS

This chapter has highlighted the fact that, from a technological point of view, the industry seems to be perfectly capable of coping with the diffusion of new, solvent free formulations and the fieldwork did not reveal any differences between US and EU firms, as they all market waterbased products alongside traditional solvent based paints. The emergence and diffusion of waterbased technologies seems quite unlikely to generate an industry shake out. Incumbents seem to be well equipped, in terms of R&D, manufacturing and distribution capabilities, to pre-empt any challenge from potential entrants.

Rather than a lack of technological capabilities, customer demand is the key to some of the problems related to the diffusion of waterbased products, particularly in the decorative paints business. The issue of 'stringency' of regulation assumes here a further twist. In a sense, it is not the industry that appears to be unwilling to accept ever more stringent targets. It is rather the customer base that does not show -yet- enough interest to make the substitution of traditional paints with waterbased products worth doing on a large scale. All interviewees stressed that technology is not an issue. However, in order to achieve the technical scale necessary to drive down costs, producers should have access to distribution channels. As it happens, independent suppliers, who are generally cautious in proposing new products to extremely cost-conscious customers, often control distribution. As the decorative paints business is so sensitive to customer preferences, regulation should aim at changing demand patterns. From this aspect, US companies are enjoying the advantage of the early moves of the US legislator, as both state and federal laws have been issued, which limit the admissible VOCs content of paints. Further, US firms also rely on different customer preferences that have 'naturally' pushed them early on to develop and adopt waterbased formulations. Given their different surfaces and tastes, customers in the US have always been eager to adopt low solvent formulations.

Existing EU and national legislation impinges only upon the manufacturing process. It can be met by adopting end-of-pipe technology. Indeed, product and process innovation exhibits some degree of decoupling: VOC reducing innovations at process level do not necessarily lead to the development of VOC free products. Thus, although environmental legislation does provide a clear focus on the innovative efforts of the industry, this might not be enough.

Customers' preferences and environmental awareness are fundamental to determining the systematic adoption of existing cleaner technologies and products.

The role of demand is evidenced by the prompt adoption of waterbased paints by the international automobile industry. Here, carmakers have actively pushed to adopt low solvents formulations under pressure from both environmental regulations and cost reduction programs. Changes in the supply chain have induced paint suppliers to take full and global responsibility for the paint process. Again, US firms seem to have benefited from an early mover advantage. As the US automobile industry quicker than the EU to react to Japanese competition, US paint suppliers also were forced to react quickly. It has been noted though, that despite the obvious reduction in VOC emissions this brings, the overall environmental impact of paint shops cannot be assessed in terms of VOCs only. Other process steps involve the use of hazardous materials necessary to prepare the car body for paint application. Also, solvent based paints and cleaning solvents are still used, and the energy intensity of the process is likely to increase, as powder coatings are adopted alongside waterbased formulations.

Finally, from a more policy-oriented perspective, the appropriate object and level of regulations needs to be defined. First, most US VOC-related regulations occur at state-level. Federal legislation exists, but it is less strict and more limited in scope, as, particularly in the decorative business, climate differences and local preferences have a major impact on paint formulations. In turn, these specificities may have a significant influence on the actual product that reaches the market. In order to accommodate such differences, federal legislation must necessarily remain flexible in setting VOC targets. It therefore is debatable whether stringent, product-level legislation should be a competence of the EU or of national authorities, which are closer to the 'sticky information' embedded in the their markets. Second, despite the fact that much current legislation affects producers only, distributors play a fundamental role in determining the market viability of alternative products. Particularly in those countries where retail and trade paints are sold via large retail chains (e.g. the UK), legislation may involve these latter in an effort to promote the diffusion of clean technologies. This point is particularly important, as some of the major retailers have launched their own brand name products. By relying on the manufacturing facilities of medium sized enterprises, some retailers are now entering the market in direct competition with other producers whose products they stock on their shelves. It may be therefore important to involve retailers in the process of consultation prior to legislation, and also to consider regulating the behavior of these companies, which play such a big part in influencing customers' choices.

9. REFERENCES

Akehurst P. (1997); Cutting Down on Volatiles; *The Chemical Engineer*; **630** (6 March): 13-5.
Ashford N. A. (1993); Understanding Technological Responses of Industrial Firms to Environmental Problems: implications for government policy; in Fisher K. And J. Schot (eds.) op. cit.
Ashford N. A., C. Ayers and R. F. Stone (1985); Using Regulation to Change the Market for Innovation; *Harvard Environmental Law Review*; **9**: 419-466.
CEPE (1999), CEPE Newsletter, various issues.
European Commission (1999), Council Directive 1999/13/EC of 11 March 1999; Office Journal L 085, 29/03/1999: 1-22.
Dorfman M., W. Muir and C. Miller (1992), *Environmental Dividends: cutting more chemical waste*, New York: Inform.
European Commission (various years), Panorama of EU Industries, Eurostat.
Fisher K. And J. Schot eds. (1993); Environmental Strategies for Industry: international perspectives on research needs and policy implications; Washington DC: Island Press.
Folmer H., H. L. Gabel and H. Opschoot eds. (1995); Principles of Environmental and Resource Economics: a guide for students and decision makers. Aldershot: Edward Elgar.
Giordan B. (1995), Will the Coatings Industry Get the Regulation It Deserves?, The Future of Industrial Coatings, 15th International Conference, Brussels, Belgium.
Groenewegen P. and P. Vergragt, (1991), Environmental Issues as Threats and Opportunities for Technological Innovation, *Technological Analysis and Strategic Management*, **3** (1), 43-55.
Howard J. P. (1999); A Colourful Prospect; *Chemistry and Industry*; nr. 10: 380-2.
Karadeloglou P., T. Ikvue, and J. Skea (1995); Environmental Policy in the European Union; in Folmer H., H. L. Gabel and H. Opschoot eds., op. cit.
Kemp R. (1993); An Economic Analysis of Cleaner Technology: theory and evidence; in Fisher K. And J. Schot (eds.) op. cit.
PRA (Paints Research Association) (2001), Coatings Comet, various issues.
Schmidheiny S. (1992), *Changing Course: a global business perspective on development and the environment*, Business Council for Sustainable Development, Cambridge: MIT Press.
Rest K. M. and N. A. Ashford (1988); Regulation and Technological Change: the case of occupational exposure to formaldehyde; *Harvard Journal of Law and Technology*; **1**: 63-96.
Skea J. (1995); Environmental Technology; in Folmer H., H. L. Gabel and H. Opschoot eds. Op. cit.
von Hippel E. (1995);"Sticky information" and the Locus of Problem Solving: implications for innovation, *Management Science*, **40**: 429-439.
Walker A. (1999); Flying Colours; *The Chemical Engineer*; **672** (14 January): 15-9.
WZB (1996), Innovation in the European Chemical Industry, EIMS Studies no. 38.

Chapter 5

RESEARCH COLLABORATIONS AMONG INVENTORS AND THE LOCATION OF R&D IN THE EUROPEAN CHEMICAL INDUSTRY

Myriam Mariani
University of Camerino and MERIT, University of Maastricht

Abstract: This work traces the geographical distribution of R&D activities in Europe in 5 chemical sectors – i.e. biotechnology, pharmaceuticals, organic chemistry, materials and polymers – and explores research collaborations at the level of the individual inventors. It compares the firm and the geographical cluster as two organisational modes for producing interdisciplinary patents and for pulling together large teams of inventors. The results confirm that chemical companies perform most of their research in the home-country, and that patenting activity clusters geographically. However, compared to the geographical cluster, the large firm is a more effective mechanism for producing interdisciplinary innovations and for fostering the formation of large and de-localised teams of inventors.

Key words: Multinational Firms; Research Collaborations; Innovation; Chemical Industry

1. INTRODUCTION

There is a long literature on the geographical distribution of production and innovative activities. Some contributions describe their location across countries and regions in Europe and the US, and analyse the implications in terms of country and regional competitiveness and growth. Others investigate the motives that lead to the geographical concentration of such activities. In the case of innovation, in particular, they argue that economies of scale in R&D

F. Cesaroni et al. (eds.), R&D, Innovation and Competitiveness in the European Chemical Industry, 119-143.

stimulate firms to locate in one geographical area. In addition, innovative activities cluster geographically because they benefit from localised technological spillovers (e.g. Jaffe, 1986; Jaffe, Trajtenberg and Henderson, 1993; Coe and Helpman, 1995; Audrestch and Feldman, 1996; Swann et al. 1998). This chapter describes the geographical distribution of R&D activities in the chemical and pharmaceutical sectors in Europe, and analyses the extent to which they cluster regionally. It also explores a new issue in the literature: it compares the relative effectiveness of the geographical cluster and the firm as organisational modes for fostering collaborations among individual inventors and for producing interdisciplinary or "general" innovations.

The empirical investigation is based on a random sample of 10,000 European chemical and pharmaceutical patents applied for at the EPO (European Patent Office, 1998) in 1986-1997. Each patent is allocated to a specific branch of the chemical industry: biotechnology, materials, organic chemistry, pharmaceuticals and polymers. I use the zip code contained in the address of the inventors to assign each patent to the European region in which it was invented. Information on the characteristics of the innovations, of the regions in which these innovations are invented, and of the firms that developed them are collected from the EPO database and other data sources.

The results confirm that also in chemicals innovative activity concentrates geographically across European regions, and that chemical companies perform most of their research in the home country. Moreover, when the geographical cluster is compared to the firm for organising R&D activities, the large company is a more effective coordination mechanism to enhance the formation of international networks of inventors, to induce a large number of inventors to collaborate, and to produce interdisciplinary patents.

The next section introduces the key issues of the chapter. Section 3 describes the geographical distribution of chemical innovations in Europe and presents some insights from the data about the effectiveness of the firm and the geographical cluster. Section 4 checks for multiple correlations by means of Probit and OLS regressions. The final section concludes.

2. ORGANISATIONAL VS. GEOGRAPHICAL PROXIMITY

Several contributions explore the motives and the benefits that arise when scientists collaborate. (See, among others, de Solla Price 1966; Merton 1973; Narin and Withlow, 1990). Moreover, when the mechanisms that give rise to research collaborations are analysed, the firm emerges as the natural and efficient mechanism to foster and coordinate research activities among individuals (Nelson and Winter, 1982; Dosi et al. 1988; Patel and Pavitt, 1997). This is because the firm develops specific competencies and learning processes, unique internal communication systems and routines that help reduce the cost of coordinating different parts of the organisation and different individuals involved in a common R&D project. Klepper and Sleeper (2002) highlight the importance of firm distinctive capabilities for spin-off companies in the automobile industry, and show that they originate from the large parent companies. They also argue that these capabilities are difficult to reproduce without transferring the human capital employed by the companies themselves. These firm specific competencies help explain the persistent different performances of firms in terms of innovativeness, competitiveness and growth (Nelson, 1995; Teece et al. 1997).

Recently, an alternative candidate to the firm as a means to coordinate inventors and research activities has emerged in the literature: this is the geographical cluster. The "economic geography" literature shows that R&D activities concentrate geographically. The characteristics of knowledge and the characteristics of the innovative process, which is complex and uncertain, help explain the motives for the geographical concentration of research activities. Complexity and uncertainty can indeed be reduced by accessing information and by fostering communication among people. Communication, in turn, is negatively influenced by the distance among the relevant agents. This is particularly so when knowledge – especially "new knowledge which tends to be informal and uncodified" (Pavitt, 1987) – is involved in the production of other knowledge or information, and when it relies on practice and learning-by-doing. Moreover, knowledge is cumulative in the sense that new knowledge builds upon previous knowledge. At the geographical level this explains why scientific and technological progress is faster in regions that have accumulated high levels of innovative activities over time.

There is also empirical evidence showing that knowledge exchange among people and institutions benefits from being geographically close. Audretsch and Feldman (1996) found that there are sectoral differences in the geographical agglomeration of innovative activities, with some industries like computer and pharmaceuticals displaying a higher degree of concentration compared to all manufacturing. Breschi (1999) derived similar conclusions by examining patent data from the European Patent Office during 1978-1991. Other contributions try to measure the geographical extent of knowledge spillovers, and study the effect of spillovers on regional economic growth (Caniels, 1999). Finally, some recent studies suggest that localised knowledge spillovers do not occur unintentionally. Zucker, Darby and Armstrong (1998) show that in California the geographical proximity of biotechnology companies with academic scientists does not stimulate per sé the economic performance of the companies. Only when these firms have "formal" relationships with the scientists in the academy (e.g. joint papers, presence in the Board of Directors), these "formalised" spillovers have a positive impact on their growth. The study on the glass industry by Lamoreaux and Sokoloff (1998) shows that in the late nineteenth and early twentieth centuries inventors were more fertile where information flows were organised by localised networks of institutions that mobilised information about technological opportunities, and mediated relations among the inventors and the users of the innovations. (See also Arora, Gambardella, and Fosfuri, 2001).

This chapter tries to reconcile these two streams of the literature. After describing the location and degree of geographical concentration of research in the chemical industry in Europe, it compares the geographical proximity to the organisational proximity among inventors as alternative mechanisms for exchanging knowledge. There is geographical proximity if the inventors are located in the same region at the NUTS2 and NUTS3 level; there is organisational proximity when the inventors are affiliated to the same firm. This work will estimate the relative importance of these two organisation modes on the probability that different types of research activities are performed: research that leads to general vs. more specific innovations, or research that involves sizeable vs. small networks of researchers.

The expectations are that:

a)Organisational proximity is a more effective mechanism for organising research collaborations compared to geographical proximity. This is because the routines and the internal coordination systems of the firm reduce the time and costs of pulling together a large number of researchers, and make it easier to coordinate complementary competencies needed in interdisciplinary projects. Therefore, patents developed by the firm are expected to have a larger number of inventors and to be more complex compared to those invented by being geographically close. This is expected to be even more so for patents developed by large multinational companies that can coordinate a larger number of specialised and complementary competencies. In this case, however, the cost of coordinating inventors located in different subsidiaries and different geographical regions might partly compensate the expected benefits.

b) Geographical proximity is a good coordination mechanism if the inventors are located in a technology intensive region – i.e. in a technological cluster. I therefore expect companies to produce more complex patents in a technological cluster than in a non-cluster area. The underlining argument is that in technology-intensive regions, where innovative activities agglomerate, it is easier to find the complementary competencies needed in complex R&D projects. Moreover, since people with complementary expertise are located very close to one another, the probability that they collaborate is higher than in non-cluster areas. I also expect this probability to be higher for smaller firms that might use the advantages of the technological cluster to compensate the lack of diversified internal competencies.

To test these propositions I selected 4,650 patents where at least one inventor is located in Europe out of a random sample of 10,000 chemical patents applied at the EPO in 1986-1997, and defined three patent characteristics:

1. Co-localised and De-localised patents (hereafter CL or DL patents). At the regional level, a patent is CL if all the inventors are located in the same NUTS2 or NUTS3 region. If one or more inventors are located in a different region, the patent is DL. I assume that a DL patent is produced by the coordination played by the organisational proximity. The logic behind this assumption is the following. Let us consider a CL patent. This is produced by inventors located geographically close. However, the information available in the patent document does not allow me to distinguish between CL patents developed by inventors belonging to the same laboratory – and hence to the same firm – or to different firms located in the same region. In other words, a CL patent can be the outcome either of the coordination played by the firm or the cluster. By contrast, a DL patent is the result of the collaboration among inventors located in different places. This means that it cannot be the outcome of the geographical proximity among inventors. As a first approximation, then, one can assume that they are the outcome of the coordination played by the firm, which is supported by the fact that 92.4% of DL patents are applied by single applicants.

2. Interdisciplinary of patents. The interdisciplinarity of a patent is proxied by the number of supplementary classes in which the patent is classified by the experts of the EPO. Each patent is classified in one main obligatory technological class according to the International Patent Classification (IPC), and if the patent officers believe that the patent falls into other technological classes as well, they assign it to other supplementary IPC classes. Intuitively, the higher is the number of supplementary classes, the more the patent is interdisciplinary and "general". In turn, more interdisciplinary patents are expected to require a higher coordination effort in order to pull together and coordinate heterogeneous expertise.

3. Breadth of the network. The breadth of the network of inventors is proxied by the number of inventors listed in the patents. Again, a higher

coordination effort is needed when a larger number of inventors are involved in a research project.

3. INSIGHTS FROM THE DATA

The data used in this chapter are drawn from different sources. I started by selecting a random sample of 10,000 chemical patents from a database of 201,531 chemical patents granted and applied at the European Patent Office (EPO, 1998) between 1986-1997. I did not make any distinction between patent applications that have been granted, and those that have not been granted yet.

By using the information on the technological class assigned by the EPO (3-digit IPC classes) and by reading the abstract of the patent, an expert pharmacologyst helped me classify each patent in one of the following 5 chemical sub-sectors: biotechnology, materials, organic chemistry, pharmaceuticals and polymers. From the EPO database I also collected information on the number, names and addresses of the applicants of each patent; the number, names and addresses of the inventors; the date of the patent application; the obligatory IPC class, and the number and type of supplementary IPC classes. The names of the applicants of these patents were standardised in order to merge mother and daughter firms under the same name. The Who Owns Whom (1995) database was used to investigate these mother-daughter relations. The Fortune 500 list (1995) was used to mark the largest applicant firms in the sample as the "Fortune 500" firms.

Finally, each innovation was located geographically. The zip code contained in the addresses of the inventors was considered as a proxy for the location of the inventive activity that led to the patent. I used it to assign each patent to the specific NUTS3, NUTS2, and NUTS1 region in which it was invented (Nomenclature des Unités Territoriales Statistiques), and to classify the patents as CL or DL. Data about the characteristics of the regions in which the innovations were developed such as the GDP, the population and the size of the regions were drawn from the EUROSTAT REGIO database (1999). From the European R&D database (Reed Elsevier Publisher, 1996) I downloaded about 9,000 chemical laboratories, classified them as private labs or public labs, and assigned to their specific NUTS regions. This part of the work was confined to a sub-sample of 4,650 patents where at least one inventor from the inventors' list had a European address.

3.1 The geographical distribution of chemical and pharmaceutical research in Europe

The largest share of the patents in the sample is invented in the USA (34.5%). Japan and Germany follow with almost half of the share of the USA (18.3% and 18.9%). Much lower shares of patents are invented in France, Great Britain, Italy, Switzerland, and in the Netherlands (6.9%, 6.7%, 3.2%, 2.4% and 2.2%). Table 1 calculates the standardised Revealed Technological Advantages of each country in the 5 technological classes (Soete, 1989). The RTA of a country is the country's share of all patenting in a given technology relative to its share of all patenting in all technologies. The standardised RTA index (RTA-1/RTA+1) varies between –1 (despecialisation) and 1 (specialisation).

Table 1: Standardised Revealed Technological Advantages of countries in biotechnology, materials, organic chemistry, pharmaceuticals and polymers (1986-1997)

(RTA-1)/(RTA+1)

Country	Biotech	Materials	Organic chemistry	Pharma	Polymers
DE	-0.3	0.0	0.1	0.0	0.1
FR	-0.1	0.0	0.0	0.1	-0.1
UK	0.0	-0.1	0.1	0.1	-0.4
IT	-0.3	-0.3	0.1	0.1	0.0
CH	-0.3	-0.6	0.2	0.0	-0.2
NL	0.1	0.3	-0.1	-0.1	0.1
BE	0.0	0.2	-0.2	0.0	0.2
SE	0.4	-0.2	-0.4	0.2	-0.6
DK	0.5	-0.2	-0.1	-0.3	-1.0
ES	0.2	-0.5	0.1	0.1	-0.4
AT	0.4	0.3	-0.6	0.0	-0.1
FI	0.3	0.0	-0.4	0.1	0.0
NO	0.4	0.0	-0.1	-0.2	-0.5
GR	0.6	-1.0	-1.0	-0.1	-1.0
LU	-1.0	-1.0	-1.0	-1.0	0.6
US	0.1	-0.1	-0.1	0.0	0.0
JP	-0.1	0.1	0.0	-0.1	0.2
Others	0.2	-0.1	0.0	0.1	-0.3

Source: EPO, 1998 Note: 10,000 observations

Table 1 shows that smaller countries (or countries that apply for a low number of chemical patents) are comparatively more technological specialised compared to countries with many patents in the chemical industry. Italy, Switzerland, Netherlands, Sweden, Denmark, Spain, Austria, Finland, Norway and Greece belong to the first group of countries. All other countries tend to technologically differentiate in the 5 chemical sub-sectors (the RTA index takes the values 0, 0.1, and – 0.1 in most of the cases).

Table 2 looks at the regional distribution of the 10,000 patents, and lists the top 20 regions in which patents are invented among the 158 European regions used in this chapter (See Appendix 1). Consistently with other works (Paci and Usai, 2000; Caniels, 1999) the distribution of chemical patents across European regions is highly concentrated. The top 20 regions account for 77.5% of the total number of patents invented in Europe. The top 10 regions invent 58.7% of the patents.

Table 2: Cumulative share and Herfindhal concentration index of chemical patents invented in the European regions (1986-1997). Top 20 regions

NUTS	Cum frequency
Nordrhein-Westfalen (DEA)	13.7
Hessen (DE7)	20.8
Île de France (FR1)	27.7
Rheinland-Pfalz (DEB)	34.3
South East Egland (UKJ)	40.5
Switzerland (CH)	45.8
Lombardia (IT2)	49.3
Baden-Württemberg (DE1)	52.6
West-Nederland (NL3)	55.7
Sachsen-Anhalt (DEE)	58.7
North West England incl. Merseyside (UKD)	61.7
Bayern (DE2)	64.6
Rhône-Alpes (FR71)	67.2
Vlaams Gewest (BE2)	69.3
Sachsen (DED)	71.1
Eastern England (UKH)	72.7
Berlin (DE3)	74.0
Niedersachsen (DE9)	75.3
Alsace (FR42)	76.4
Zuid-Nederland (NL4)	77.5
Herfindhal	0.05

Source: EPO, 1998 Note: 10,000 observations

Nordrhein-Westfalen, Hessen and Paris are in the top positions. There are 9 German regions among the top 20, and 52% of the patents invented in the top 20 regions are invented in Germany. France (13.8%), the UK (13.8%), the Netherlands (5.3%) and Italy (4.6%) follow Germany in the ranking in Table 2. Greece, Spain, the South of Italy, Portugal, Sweden and some Danish and Finnish regions are at the bottom of the list. Moreover, since European regions differ in size, I calculated the number of per capita patents in each region (not shown here). Although the ranking is slightly different, 17 of the 20 top regions in terms of per capita patent activity are also in the top 20 positions in Table 2. The 3 missing regions still perform high per capita patenting activity.

Finally, I computed the regional cumulative distribution of patents and the Herfindhal indexes for each of the 5 sectors.

Table 3: Cumulative share and Herfindhal concentration index of patents invented in the European regions in biotechnology, materials, organic chemistry, pharmaceuticals and polymers (1986-1997). Top 20 regions

Biotech		Materials		Organic chem.		Pharmaceuticals		Polymers	
UKJ	8.2	DEA	14.5	DEA	15.3	DEA	12.5	DEA	20.1
FR1	15.6	DE7	22.3	DE7	24.5	FR1	22.9	DEB	31.9
DE2	20.7	FR1	28.8	DEB	32.9	UKJ	30.8	DE7	38.3
DE7	25.7	NL3	34.0	CH	40.7	DE7	36.6	CH	42.2
NL3	30.6	DEB	39.1	UKJ	47.7	CH	41.3	FR71	46.0
CH	34.6	DE1	43.5	FR1	53.9	UKD	45.9	IT2	49.6
UKH	38.4	UKD	47.6	DEE	58.0	IT2	50.2	DE1	52.9
DK002	41.6	BE2	51.6	IT2	61.9	DEB	53.9	FR1	55.9
DEA	44.7	DED	55.3	DE1	65.4	DE1	56.9	DED	58.8
DE9	47.6	DE2	58.5	FR71	68.3	NL3	59.9	DEE	61.5
DE1	50.3	DE9	61.3	UKD	71.0	DEE	62.8	NL3	64.1
BE2	52.8	FR71	64.0	DE2	73.4	DE2	65.5	DE2	66.7
FR71	55.1	NL4	66.7	NL3	75.6	DE3	67.6	NL4	69.1
AT1	57.2	UKC	69.3	BE2	77.3	BE2	69.4	BE2	71.5
DE3	59.1	UKJ	71.5	DED	79.0	IT6	71.2	UKD	73.8
IT2	60.9	DEE	73.1	FR42	80.3	FR71	72.9	IT4	75.9
DEE	62.6	FR42	74.5	UKH	81.6	UKH	74.6	BE3	77.9
FR42	64.3	FR82	75.8	DE3	82.8	DED	76.0	UKJ	79.7
BE3	65.9	NL2	77.2	UKM	83.9	UKC	77.1	DE9	81.2
FI161	67.5	AT1	78.5	IT11	84.7	UKF	78.2	BE1	82.5
Herf.	0.03		0.05		0.06		0.05		0.07

Source: EPO, 1998

Note: 10,000 observations

Table 3 lists the top 20 regions in which patents in biotechnology, pharmaceuticals, organic chemistry, polymers and materials are invented. The concentration indexes indicate that patenting activity concentrates geographically in all 5 chemical sectors. Still, there are differences across technologies. Polymer and organic chemical patents cluster more, while biotechnology patents display the lowest concentration index. Moreover, most of the regions appear on all 5 listings in the top 20 positions. This is the case, for example, of Vlaams Gewest, Switzerland, Darmstadt, Ile de France, Rhone-Alpes, West-Nederland, and South East England. Although their ranking differs according to the sector, these regions are the most active in all 5 branches of the chemical industry. Differently, in biotechnology there are some "new entry" regions like Frederiksborg amt in Denmark, Uusimaa in Finland and Stockholm (which is ranked 22nd in biotechnology) that appear in the top 20 positions in biotechnology while performing far behind in the other 4 chemical fields.

3.2 Home-country vs. foreign location

Traditional theories suggest that research and innovation are located in the companies' home country that pursues competitive frontiers in new product markets while transferring more mature technologies abroad. This is the basic theory developed in the product cycle model of Vernon (1966). It also corresponds to the diffusion of the industrial revolution from England to other countries during nineteenth century. In both cases entry into new geographical markets was motivated by the opportunity to exploit economies of scale and scope based upon technologies pioneered in the home market. (Chandler, 1990).

Some recent work, however, points out that multinational firms are increasingly moving research to foreign locations (Cantwell, 1994; Cantwell and Janne, 1998). From the US perspective, Arora and Gambardella (1998) find that company financed R&D performed outside the USA by American companies in the chemical and pharmaceutical sectors rose from roughly 13% in 1985 to 25% in 1995. Walsh (1997) listed the 50 firms with the largest share of patents coming from outside their country. Eight of them are chemical companies, and three are in pharmaceuticals.

To examine the extent to which research in the chemical industry is located abroad, this section compares the location of the applicants of the 10,000 patents in the sample with the location of their inventors. The location of the inventors of the patent is thought as the locus where the innovative activity takes place. The location of the applicant is given by the nationality of the mother company of the firm that applied for the patent. The share of patents attributable to inventors located in the home country of the applicant is 88.4%. Only 11.6% of the total patent sample is invented in a foreign country, and this share does not change over the period 1986-1997. Therefore, chemical companies continue to perform a high proportion of innovative activity in their home country. Table 4 shows the location of R&D from the perspective of the major world regions.

Table 4: Share of patents invented by European, American and Japanese applicants in Europe, USA and Japan (1986-1997)

	Region of the applicant				
Region of the invention	EU applicants	US applicants	JP applicants	OTH applicants	Total
EU	86.0%	9.0%	1.0%	11.5%	44.8%
US	12.2%	87.7%	2.3%	5.3%	34.5%
JP	0.9%	2.4%	96.5%	0.1%	18.3%
OTH	0.9%	0.9%	0.2%	83.1%	2.4%
Total	100%	100%	100%	100%	100%

Source: EPO, 1998

Note: 10,000 observations

European companies locate in Europe 86.0% of their research in chemicals. The share of US patents developed in the US is 87.7%. It increases to 96.5% for Japanese companies. When European chemical companies locate R&D outside Europe, they develop about half of their patents in the US, and the other half in other countries. The American favourite foreign location is Europe: 12% of total American patents in chemicals are invented in Europe (62% of patents applied overseas by US companies in chemicals). Table 5 shows the share of patents invented in the home country and in foreign locations by European, American and Japanese companies in the 5 chemical sectors.

Table 5: Share of patents invented by European, American and Japanese applicants in Europe, USA and Japan (1986-1997). Sectoral breakdown

	Region of the applicant			
Region of the invention	EU applicants	US applicants	JP applicants	OTH applicants
Biotechnology				
EU	80.2%	0.7%	5.1%	36.5%
JP	1.4%	94.9%	1.0%	14.6%
OTH	1.8%	0.0%	1.4%	3.5%
US	16.6%	4.4%	92.5%	45.4%
Total	100.0%	100.0%	100.0%	100.0%
Materials				
EU	90.7%	0.9%	8.0%	44.9%
JP	0.9%	98.4%	1.4%	22.2%
OTH	0.7%	0.4%	0.6%	2.0%
US	7.8%	0.2%	90.1%	30.9%
Total	100.0%	100.0%	100.0%	100.0%
Organic chemistry				
EU	89.1%	0.7%	10.8%	50.8%
JP	0.8%	97.5%	1.3%	18.7%
OTH	0.6%	0.4%	0.5%	2.2%
US	9.5%	1.4%	87.4%	28.4%
Total	100.0%	100.0%	100.0%	100.0%
Pharmaceuticals				
EU	85.0%	2.6%	11.5%	47.3%
JP	0.8%	95.2%	1.3%	13.9%
OTH	0.9%	0.0%	1.1%	2.7%
US	13.3%	2.2%	86.2%	36.0%
Total	100.0%	100.0%	100.0%	100.0%
Polymers				
EU	85.4%	0.4%	8.3%	40.1%
JP	0.8%	96.6%	6.6%	25.1%
OTH	0.9%	0.1%	0.5%	1.4%
US	12.9%	2.9%	84.6%	33.5%
Total	100.0%	100.0%	100.0%	100.0%

Source: EPO, 1998

Note: 10,000 observations

The applicant companies tend to perform research at home in all 5 sectors. The only exception is biotechnology. Specifically, European companies invent 80.2% of biotechnology patents in Europe, and almost 16.6% in the US. The share of biotechnology patents invented abroad falls to 4.4% for Japanese companies and 7.5% (5.1% in Europe) for American companies. Hence, in biotechnology, the newest and more science-intensive branch of the industry, European companies locate research in the USA more than in the other sectors, and more than other countries do. The "attractiveness" of the USA as a locus for doing research in biotechnology is consistent with the fact that the USA has probably developed strong technological competencies in this field: 45.4% of biotechnological patents in the sample are invented in the US. Only 36.5% are in Europe, and 14.6% in Japan.

3.3 Geographical vs. organisational proximity as coordination mechanisms for research collaborations. A first look.

The empirical investigation in this section uses a sample of 4,650 patents applied for in 1986-1997 where at least one inventor is located in Europe. I checked several features of the 4,650 patents, and they appear to be fairly representative of the whole population.

Most of the patents (92.4%) are applied by single applicants. Only 353 patents are applied by multiple applicants (up to 6). The analysis below is based on the first applicant. Their composition is as follows: 159 are Public institutions, excluding universities; 44 are universities; 133 are individual inventors (i.e. applicants who do not belong to any firm); 2,123 are large firms listed in Fortune 500 (1997) – i.e. the "Fortune 500" firms. The remaining 2,191 firms are not listed in Fortune 500 (1997). I will term them as the "other" firms. Moreover, 788 patents (16.9%) are developed by single inventors, while 83.1% are invented by multiple inventors.

To check whether organisational proximity is a better mechanism for fostering research collaborations compared to the geographical proximity among inventors, I started by computing the share of DL patents over the total number of patents in the sample. Thirty eight per cent of total patents are DL. The share of CL patents is 62%. The higher share of CL compared to DL patents is suggestive of the fact that the geographical proximity among inventors plays a role in pulling them together. Unfortunately, I cannot distinguish whether this proximity is either among inventors belonging to the same firm, or among inventors located in the same geographical area and belonging to different firms.

I then compared the characteristics of DL and CL patents. The results (not shown here) reveal that the teams of inventors that produce DL patents (4.1 inventors on average per patent), and the interdisciplinarity of these innovations (2.1 IPC supplementary classes on average) are greater than those for the whole sample and, clearly, for CL patents (2.7 inventors on average and 1.8 IPC supplementary classes). The low share of DL patents in the sample, however, suggests that the vast majority of companies do not have internally the competencies to coordinate such collaborations across distance. To understand whether large firms are better than smaller firms at producing these DL patents Table 6 shows the share and the average characteristics of patents developed by the "Fortune 500" firms, and the difference between these means and those of patents developed by the "other" firms.

Table 6: "Fortune 500" firms vs. "other" firms: Share of DL patents, mean number of inventors and mean number of supplementary classes by patent. Differences between means

	Share of DL patents over the total number of patents produced by the two types of companies	Number of inventors: Mean and differences between means	Number of supplementary classes: Mean and differences between means
"Fortune 500" firms	900 *41.8%*	3.5 (0.04)	2.1 (0.04)
"Others"	771 *30.9%*	-0.6 (0.06)	-0.3 (0.06)
Total	1671 *38.7%*	3.2 (0.03)	2.0 (0.03)

Source: EPO (1998) and Fortune 500 list (1995)
Note: Standard Errors in parenthesis. These data do not include universities, government and individual inventors. The number of observations is 4,314.

The results indicate that the "Fortune 500" companies produce a higher percentage of DL patents than the "other" firms. The share of DL patents over the total number of patents invented by the "Fortune 500" firms is 41.8%, compared to 30.9% of DL patents produced by the "others". Patents applied by these multinational companies are also produced by a larger number of inventors (3.5 inventors on average) and are more interdisciplinary (2.1 supplementary classes on average) than patents produced by the "other" firms. On average, the "other" firms' patents are invented by 2.9 inventors, and list 1.8 supplementary classes. Hence, large companies coordinate larger networks of inventors to produce more interdisciplinary patents than the "others". Moreover, when patents invented by the "Fortune 500" companies are DL, the networks of inventors become even larger. The average number of inventors listed in DL patents produced by the "Fortune 500" companies is 4.3, and the number of supplementary classes is 2.2 (not shown here). These differences are significant.

The second expectation was about the effect of the technological characteristics of the regions in which the inventors are located over the probability that the inventors collaborate, and on the characteristics of the outcome patents. Specifically, the expectation was that geographical proximity in technology-intensive regions gives rise to more interdisciplinary patents and to larger networks of inventors compared to non-cluster regions.

I started by defining a technological cluster. The EPO database (1998), the Eurostat Regio database (1995) and the European R&D database (1996) provide information on the regional characteristics of the European regions such as the GDP, the extension of the geographical area, the number of employees in the manufacturing and service sectors, the population, the patents invented and applied in the regions in different sectors, and the number of chemical laboratories established by public and private institutions. I examined the distribution of these variables across the European regions, and decided to use the number of chemical laboratories – both private and public – as a proxy for the technological intensity and infrastructure developed by a region in the chemical sector. The number of chemical laboratories ranges between 0 and 647, and their distribution is very skewed: 67 out of 91 regions host less than 100 laboratories. Only the last quartile of the European regions in the sample shows between 100 and 647 chemical laboratories. The regions in this quartile are the "technological clusters". As expected, other variables, such as the number of patents invented in each area are correlated with the number of laboratories. I also checked that the firms in the database did not determine themselves the characteristics of the cluster. I calculated the share of laboratories owned by the applicants of the patents over the total number of laboratories in each region. The results show that these companies own only a small fraction of the total number of laboratories in each region.

Table 7 shows the probability of a patent being CL in a cluster region and compares it to the probability of a patent being CL in a non-cluster region. This probability goes from 66% in the "clusters" to 48.3% in non-cluster areas. The share of DL patents increases correspondingly. This suggests that the location in a cluster may play a role in pulling researchers together.

Table 7: Cluster vs. non-cluster regions: Share of CL patents, mean number of inventors and mean number of supplementary classes by patent. Differences between means

	Share of CL patents over the total number of patents produced in the two types of regions	Number of inventors: Mean and differences between means	Number of supplementary classes: Mean and differences between means
Cluster regions	1,816 *66.1%*	3.3 (0.05)	1.9 (0.05)
Non-cluster regions	578 *48.3%*	-0.2 (0.06)	0.1 (0.06)
Total	3,944 *60.7%*	3.2 (0.03)	2.0 (0.03)

Source: EPO (1998) and European R&D database (1996)
Note: Standard Errors in parenthesis. Note: these data do not include universities, government and individual inventors. The number of observations is 3,944.

Table 7 also shows that the mean number of inventors is significantly higher for patents invented in the cluster regions. The number of supplementary classes is instead slightly higher in non-cluster regions. These results, however, might be due to the fact that the table does not highlight the net effect of being in a cluster over the variables of interests. The multiple correlation analysis in the next section will check for this.

Finally, I tried to understand who benefits more – i.e. large vs. small firms – from being located in the clusters. The technological cluster, like the large company, typically features a good deal of different and complementary competencies inside the territorial area. Therefore, firms located in a technological cluster have limited need to find these competencies outside the region. The probability of a patent being produced by "others" is 42.0% in non-cluster areas and rises to 50.0% in the clusters (not shown here). Symmetrically, the probability of a patent being produced by a "Fortune 500" firm decreases from 58.0% in non-cluster areas to 50.0% in the clusters. These results support the idea that the technological cluster is comparatively more attractive for smaller and less global firms compared to large multinational companies. Although both small and large firms take advantage from being in a technological cluster, geographical proximity in a technological intensive region seems to be more important for companies that lack the internal scientific competencies and the organisation capabilities to coordinate the collaborations. These results also confirm the idea that the "Fortune 500" firms are comparatively better at coordinating competencies and researchers located in different parts of the organisation and in different geographical areas.

4. MULTIPLE CORRELATIONS: PROBIT AND OLS REGRESSIONS.

The tables in the previous section show two interesting results. First, both the "Fortune 500" companies and the "other" firms develop more CL patents than DL patents. This suggests that geographical proximity matters also for the large multinational firms. They develop innovations in specific regions or subsidiaries by pulling together competencies locally, at the regional level. However, the higher share of DL patents invented by the "Fortune 500" companies compared to the "other" firms, and the higher average number of inventors and supplementary classes listed in their patents suggest that large companies may also act as global networks for the development of innovations. When thy coordinate inventors located in different places, the organisational proximity among inventors produces more interdisciplinary patents with larger networks of inventors compared to the geographical proximity. Second, being in a technological cluster positively affects the probability of the inventors to be co-localised. It is not clear, however, what is the sign of the relationship between the location in a technological cluster and the other patent characteristics. More generally, some of the results shown in the tables are inconclusive because they cannot isolate the net effect of the independent variables over the probability of a patent having certain characteristics. To do so, this section performs a multiple correlation exercise by means of PROBIT and OLS regressions. The purpose of these regressions is to describe the effect of regional and firms' characteristics over the probability of a patent being co-localised, over the number of inventors that invent a patent, and over the probability of a patent being interdisciplinary.

The total sample is composed of 4,650 European patents. I run a Probit regression that estimates the probability of a patent being CL or DL, and two OLS regressions in which the number of inventors that take part in the development of a patent (NRINV) and the number of supplementary classes (NRCL) are the dependent variables. The regressors are country, sectoral, regional and firms' characteristics. Table 8 lists the variables used in the regressions.

Table 8: List of variables

CL and DL	Dependent variable of the Probit regression: it takes the value 1 when the inventors are co-localised; 0 otherwise
NRINV	Dependent variable of the OLS regression: number of inventors that collaborate in a patent
NRCL	Dependent variable OLS regression: number of supplementary classes in a patent
LABS	Regional density of chemical R&D laboratories. This is given by the number of chemical R&D laboratories in the NUTS2 European regions (stock in 1996) over the geographical extension of the regions (in Km^2)
BR	Regional density of patents invented in all sectors. This is calculated as the number of patents in all sectors invented in the NUTS2 European regions (average 1986-1990) over the population of the regions (average 1986-1990)
GDP	Gross Domestic Product per capita, average 1986-1994 in PPP and corrected for inflation
DENS	Regional population density (average 1986-1994)
MNE	Dummy for firm characteristics: it takes the value 1 if the applicant firm is a "Fortune 500"; 0 if it belongs to the "other" group
DSEC	Dummy for the sector to which the patent belongs: biotechnology (BIOTECH), materials (MAT), organic chemistry (ORG), pharmaceuticals (PHAR), and polymers (POL)
DCY	Dummy for the country in which the patent is invented: France (FR), UK (UK), Germany (DE), Italy (IT), Rest of Europe (OTH)
DORG	Dummy for the type of applicant: public research institute or a hospital (PUB), university (UNI), individuals (IND), firms (FIRM)

Source: EPO (1998), Fortune 500 list (1995), (Regio (1999), European R&D database (1996)

The technological characteristics of the regions are proxied by LABS and BR. LABS is the regional density of chemical laboratories in the 91 European regions where the sample patents are invented. I expect the density of chemical laboratories to have a positive impact on the probability that the inventors are co-localised, on the interdisciplinarity of the patents and on the breadth of the network of inventors. A rich external environment should offer a wide range of opportunities to find specialised and complementary competencies. BR is the ratio between the number of patents in all sectors invented in the regions over population. It complements the LABS variable as a proxy for the general technological intensity of the regions. I expect the co-localisation of the inventors, the interdisciplinary of a patent and the breadth of the network of inventors being positively influenced by a technologically conducive external environment. There are also two regional controls: GDP and DENS. GDP is the Gross Domestic Product per capita, average 1986-1994 (in PPP and corrected for inflation). The GDP per capita indicates the economic development of a region. DENS is the regional population density.

The dummy MNE takes the value 1 when the applicant is a "Fortune 500" firm, and 0 otherwise. DORG is a dummy for the type of organisation that applied for the patent: public research institute or hospital (PUB), university (UNI), individual (IND) and firm (FIRM). DSEC is a dummy for the sector of the patent: biotechnology (BIOTECH), materials (MAT), organic chemistry (ORG), pharmaceuticals (PHAR) and polymers (POL). Finally, DCY is a dummy for the country in which the patent is invented: France (FR), UK (UK), Germany (DE), Italy (IT) and Rest of Europe (OTH). All the variables are in logs. The results of the econometric estimates are shown in Table 9.

Table 9: Estimates of PROBIT and OLS regressions.

		Dependent variables		
		CL (1) vs. DL (0) patents	Number of inventors (NRINV)	Number of supplementary classes (NRCL)
β_0	Constant	7.89	0.45	0.72
		(2.10)	(0.91)	(0.97)
β_1	LABS	0.26	-0.01	-0.03
		(0.09)	(0.04)	(0.04)
β_2	BR	-0.08	-0.004	0.03
		(0.06)	(0.02)	(0.025)
β_3	GDP	-1.92	0.13	0.11
		(0.52)	(0.23)	(0.24)
β_4	MNE	-0.12	0.18	0.09
		(0.05)	(0.021)	(0.022)
β_5	DENS	0.07	0.02	0.042
		(0.09)	(0.04)	(0.05)
β_6	DUNI	0.06	0.06	-0.04
		(0.23)	(0.10)	(0.11)
β_7	DFIRM	0.38	-0.1	-0.14
		(0.12)	(0.05)	(0.06)
β_8	DIND	0.79	-0.54	-0.21
		(0.19)	(0.08)	(0.08)
Country dummies		Yes	Yes	Yes
Sector dummies		Yes	Yes	Yes
Number of observations		3,681	3,681	3,681
Log-likelihood		-2363.26	-3044.83	-3266.21

Source: EPO (1998), Fortune 500 list (1995), (Regio (1999), European R&D database (1996)
Note: Standard errors in parenthesis

As far as the probability of a patent being co-localised, the PROBIT estimation shows that both firm and regional characteristics matter, but with the opposite sign. If the inventors are located in a cluster region the probability that they are co-localised increases. More precisely, the estimated coefficient of LABS is positive and significant, while the coefficient of BR is not significant. This suggests that the specific technological environment external to the firm, which is measured by the density of the chemical laboratories in the region, positively correlates with the probability of the inventors to set up local collaborations. By contrast, once controlling for LABS, the characteristics of the more general technological environment does not add anything to the probability of a patent being CL. In other words, what matters for establishing R&D collaborations with other inventors in the a region is the presence of researchers working in the same technological field.

The estimated coefficient of MNE is negative and significant. This confirms the relationships highlighted in the tables in Section 3. If a large multinational company applies for a patent, the expected probability of the patent being DL is higher than for the "other" companies, meaning that large firms are better means for coordinating collaborations across distances.

The control variable DENS is not significant. GDP takes a negative and significant coefficient, suggesting that DL patents tend to be produced in richer regions, probably the same regions in which many large applicants are also located.

The estimates of the OLS regressions confirm the role played by the large multinational companies in developing innovations. When a patent is produced by a "Fortune 500" company, the expected breadth of the teams of inventors is larger and the degree of interdisciplinarity of the innovations is higher. The estimated coefficient of MNE is positive and significant in both regressions suggesting that large firms have the competencies and the internal coordination system to pull together a large number of inventors and to develop more complex patents as shown by the effect of MNE over the interdisciplinarity of the patents.

Nevertheless, the coordination provided by the large firms is subtle to understand. Although large companies perform a higher share of DL patents compared to the "other" companies, they still perform a high number of CL patents. This suggests that only few firms have the capabilities to coordinate R&D projects across distances, and confirms that the vast majority of large multinational companies still organise their innovative activities on local rather than on global basis (See, for example, Patel and Pavitt, 1992). When they act as "global" networks that harness expertise in different places and link them together through internal coordination mechanisms, they produce more interdisciplinary innovations and coordinate larger networks of inventors.

Contrary to expectations, once controlling for firm specific effects, the location of the inventors in a technological cluster does not affect the breadth of the networks of inventors and the degree of interdisciplinarity of the patents. The coefficient of LABS and BR are small and not significant in both OLS regressions. Hence, the extent to which chemical research clusters in the region and the quality of the general technological environment external to the firms (as proxied by the number of patents in all sectors invented in the regions) does not add anything to the probability that large teams of researchers collaborate and that "general" chemical patents are invented. The control variables GDP and DENS are also not significant.

All in all, when the firm is compared to the technological cluster as organisational modes for producing interdisciplinary innovations to which a large network of inventors collaborate, the results show that: 1) Organisational proximity in large companies is more effective than organisational proximity in smaller companies for coordinating inventors across distances. By contrast, the technological environment developed in the chemical sector in the regions better acts as a vehicle for inducing locally-bounded collaborations; 2) Organisational proximity in large companies is also a better coordination mode for pulling together larger networks of inventors and for producing more "general" patents compared to the cluster; 3) Geographical proximity in a technological cluster does not affect the probability of a patent being interdisciplinary and invented by a large team of inventors.

5. CONCLUSIONS

This work describes the spatial distribution of innovative activities in Europe in the chemical sector. It also compares the role of the geographical proximity among inventors in a technological cluster to their affiliation to the same chemical company (organisational proximity) for establishing research collaboration among inventors and for producing interdisciplinary patents. To do so it uses a sample of 10,000 chemical patents in 1986-1997, and firm and regional level information drawn from the "European R&D database", Eurostat Regio, and other sources. The geographical unit of analysis is the European regions according to the NUTS classification at the second and third level of disaggregation.

The results confirm that inventive activity concentrates geographically across European regions also in the chemical industry. The top 20 regions account for 77.5% of the total number of patents invented in Europe. The top 10 regions invent 58.7% of these patents. Moreover, many of the top 20 regions in which patents are invented are the same across 4 or 5 sectors. Only biotechnology has potentialities for regions that are not traditionally specialised in developing chemical innovations.

Chemical companies also perform little research outside their home country. The overall average of overseas invented patents is 11.6%. European companies locate 86.0% of their research activity in chemicals in Europe. In biotechnology, this share drops to 80.2% suggesting that European companies locate biotechnology research abroad more than in the other chemical sectors and much more than the other countries do. The share of biotechnology patents invented abroad by European companies is 19.8% (16.6% invented in the US vs. 5.1% of American biotechnological patents invented in Europe).

As far as the organisation of the innovation process is concerned, the comparison between the large firm and of geographical cluster as mechanisms for inducing collaborations among inventors shows that the larger the firm is: a) the lower is the probability that inventors are co-localised, in the sense that a lower percentage of patents is produced by inventors located in the same place; b) the larger is the network of inventors that collaborate to produce a patent; c) the higher is the number of supplementary classes listed in the patent. By contrast, the technological characteristics of the regions positively affect the probability of the inventors to be co-localised, but they show not to be correlated with the breadth of the networks and with the level of interdisciplinarity of the innovations. Hence, compared to the geographical proximity in a technological cluster, organisational proximity in large companies enhance international networks of inventors, induce a greater number of inventors to collaborate and produce more interdisciplinary patents.

6. MAIN REFERENCES

Arora A., Fosfuri A. and Gambardella A., 2001, Markets for Technology MIT Press, Cambridge (MA)

Arora A. and Gambardella A., 1998, Evolution of Industry Structure in the Chemical Industry, in Chemicals and Long-Term Economic Growth. Insights from the Chemical Industry, edited by Arora A., Landau R. and Rosenberg N.

Audretsch D. B. and Feldman M. P., 1996, R&D Spillovers and the Geography of Innovation and Production, The American Economic Review, 86 (3), 630-640

Breschi S., 1999, Spatial Patterns of Innovation: Evidence from Patent Data, in The Organisation of Innovative Activity in Europe edited by A. Gambardella and F. Malerba, Cambridge University Press, Cambridge (MA)

Caniels M., 1999, Regional Growth Differentials. The Impact of Locally Bounded Knowledge Spillovers, Ph.D Thesis, University of Maastricht

Cantwell J., 1995, The globalisation of technology: what remain of the product cycle model?, Cambridge Journal of Economics, 19, 155-74

Cantwell J. and Janne O., 1998, Globalisation of Innovatory Capacity: the Structure of Competence Accumulation in European Home and Host Countries, Working Paper, University of Reading

Chandler A., 1990, Scale and Scope, Harvard University Press, Cambridge, MA

Coe D.T. and Helpman E., 1995, International R&D Spillovers, European Economic Review, 39, 859-887

de Solla Price D.J. and Beaver D.B., 1966, Collaboration in an invisible college, American Psychologist 21 (1), 1011-1018

Dosi G. et al. (eds.), 1988, Technological Change and Economic Theory, F. Pinter, London.

Griliches Z., 1990, Patent statistics as economic indicators: a survey, Journal of Economic Literature, 28, 1661-1707

Jaffe A., 1986, Technological Opportunity and Spillovers of R&D: Evidence from Firms' Patents, Profits and Market Value, American Economic Review 76(5), 984-1001

Jaffe A., Trajtenberg M. and Henderson R., 1993, Geographic Localization of Knowledge Spillovers as Evidenced by Patent Citations, Quarterly Journal of Economics, 63 (3), 577-598

Klepper S. and Sleeper S., 2002, Entry by Spinoffs, Working Paper, Carnegie Mellon University

Lamoreaux N., and Sokoloff K., 1998, Inventors, Firms, and the Market for Technology: US Manufacturing in the Late Nineteenth and Early Twentieth Centuries, in Lamoreaux N., Raff D., and Temin P. (eds.), Learning by Firms, Organizations, and Nations

Merton R.K., 1973, The sociology of science, The University of Chicago Press, Chicago

Narin F. and Withlow E.S., 1990, Measurement of scientific cooperation and coauthorship in CEC-related areas of science, Vol. 1 Commission of the European Communities (EUR 12900 EN), Brussels

Nelson R, 1995, Recent Evolutionary Theorizing about Economic Change, Journal of Economic Literature, 33, 48-90

Nelson R. and Winter S., 1982, An Evolutionary Theory of Economic Change, Harward University Press, Cambridge (MA)

Paci R. and Usai S., 2000, Technological enclaves and industrial districts: an analysis of the regional distribution of innovative activity in Europe, Regional Studies, 34 (2), 97-114

Patel P. and Pavitt K., 1997, The Technological Competencies of the World's Largest Firms: Complex and Path Dependent, but not Much Variety, Research Policy, 26, 141-56

Pavitt K., 1987, On the nature of technology, Brighton, University of Sussex -- SPRU

Soete L., 1989, The impact of technological innovation on international trade patterns: the evidence reconsidered, Research Policy, 16

Swann P., Prevezer M. and Stout D. (eds.), 1998, The Dynamics of Industrial Clustering, Oxford University Press

Teece D.J., Pisano G. and Shuen A., 1997, Dynamic Capabilities and Strategic Management, Strategic Management Journal,18 (7), 509-533

Vernon R. , 1966, International Investment and International Trade in the Product Cycle, Quarterly Journal of Economics, 80 (2) 190-207

Walsh V., 1997, Globalisation of innovative capacity in the chemical and related products industry, Technology, innovation and competitiveness edited by Howells, J. and Michie J.

Zucker L., Darby M. and Armstrong J., 1998, Geographically localized knowledge: Spillovers or markets?, Economic Inquiry 36, 65-86

7. APPENDIX

Appendix 1. List of regions.

Region		Region	
At1	Ostösterreich	FI161	Uusimaa (suuralue)
At2	Südösterreich	FI162	Uusimaa (maakunta)
At3	Westösterreich	FI171	Etelä-Suomi
Be1	Bruxelles	FI172	Varsinais-Suomi
Be2	Vlaams Gewest	FI173	Kanta-Häme
Be3	Région Wallonne	FI174	Pirkanmaa
De1	Baden-Württemberg	FI175	Päijät-Häme
De2	Bayern	FI176	Kymenlaakso
De3	Berlin	FI177	Etelä-Karjala
De4	Brandenburg	It11	Piemonte
De5	Bremen	It12	Valle d'Aosta
De6	Hamburg	It13	Liguria
De7	Hessen	It2	Lombardia
De8	Mecklenburg-Vorpommern	It31	Trentino-Alto Adige
De9	Niedersachsen	It32	Veneto
Dea	Nordrhein-Westfalen	It33	Friuli-Venezia Giulia
Deb	Rheinland-Pfalz	It4	Emilia-Romagna
Dec	Saarland	It51	Toscana
Ded	Sachsen	It52	Umbria
Dee	Sachsen-Anhalt	It53	Marche
Def	Schleswig-Holstein	It6	Lazio
Deg	Thüringen	It71	Abruzzo
Es11	Galicia	It72	Molise
Es12	Principado de Asturias	It8	Campania
Es13	Cantabria	It91	Puglia
Es21	Pais Vasco	It92	Basilicata
Es22	Comunidad Foral de Navarra	It93	Calabria
Es23	La Rioja	Ita	Sicilia
Es24	Aragón	Itb	Sardegna
Es3	Comunidad de Madrid	Nl1	Noord-Nederland
Es41	Castilla y León	Nl2	Oost-Nederland
Es42	Castilla-la Mancha	Nl3	West-Nederland
Es43	Extremadura	Nl4	Zuid-Nederland
Es51	Cataluña	Ukc	North East
Es52	Comunidad Valenciana	Ukd	North West Eng.
Es53	Baleares	Uke	Yorkshire and The Humber
Es61	Andalucia	Ukf	East Midlands
Es62	Murcia	Ukg	West Midlands
Es63	Ceuta y Melilla	Ukh	Eastern

Es7	Canarias	Uki	London
Gr11	Anatoliki	Ukj	South East
Gr12	Kentriki Makedonia	Ukk	South West
Gr13	Dytiki Makedonia	Ukl	Wales
Gr14	Thessalia	Ukm	Scotland
Gr21	Ipeiros	Ukn	Northern Ireland
Gr22	Ionia Nisia	Fr1	Île de France
Gr23	Dytiki Ellada	Fr21	Champagne-Ardenne
Gr24	Sterea Ellada	Fr22	Picardie
Gr25	Peloponnisos	Fr23	Haute-Normandie
Gr3	Attiki	Fr24	Centre
Gr41	Voreio Aigaio	Fr25	Basse-Normandie
Gr42	Notio Aigaio	Fr26	Bourgogne
Gr43	Kriti	Fr3	Nord - Pas-de-Calais
CH	Switzerland	Fr41	Lorraine
D001	København	Fr42	Alsace
D002	Københavns amt	Fr43	Franche-Comté
D003	Frederiksborg amt	Fr51	Pays de la Loire
D004	Roskilde amt	Fr52	Bretagne
D005	Vestsjællands amt	Fr53	Poitou-Charentes
D006	Storstrøms amt	Fr61	Aquitaine
D007	Bornholms amt	Fr62	Midi-Pyrénées
D008	Fyns amt	Fr63	Limousin
D009	Sønderjyllands amt	Fr71	Rhône-Alpes
D00A	Ribe amt	Fr72	Auvergne
D00B	Vejle amt	Fr81	Languedoc-Roussillon
D00C	Ringkøbing amt	Fr82	Provence-Alpes-Côte d'Azur
D00D	Århus amt	Fr83	Corse
D00E	Viborg amt	Fr9	Overseas France
D00F	Nordjyllands amt	Pt1	Portugal Continent
FI131	Etelä-Savo	Pt2	Açores
FI132	Pohjois-Savo	Pt3	Madeira
FI133	Pohjois-Karjala	SE01	Stockholm
FI134	Kainuu	SE02	Östra Mellansverige
FI141	Väli-Suomi	SE04	Sydsverige
FI142	Keski-Suomi	SE06	Norra Mellansverige
FI143	Pohjanmaa	SE07	Mellersta Norrland
FI144	Keski-Pohjanmaa	SE08	Övre Norrland
FI151	Pohjois-Suomi	SE09	Småland med öarna
FI152	Pohjois-Pohjanmaa	SEA	Västsverige

Source: Regio (1999)

Chapter 6

COLLABORATION BETWEEN A RESEARCH UNIVERSITY AND FIRMS AND OTHER INSTITUTIONS

Aldo Geuna[*], Patrick Llerena[**], Mireille Matt[**] and Maria Savona[*]
[*]*SPRU-University of Sussex and* [**]*BETA-University of Strasbourg*

Abstract: This paper examines the evolution of contractual relationships in chemistry at the University Louis Pasteur of Strasbourg during the 1990s. Using entry, exit and persistency indicators we analyse the changes in the contracting behaviour of the various university laboratories and compare the types of contracts signed by persistent and non-persistent laboratories, persistent laboratories being those that had contractual relationships throughout the period. Four main conclusions emerge. First, an increasing number of new actors (firms and university laboratories) have become contractual partners. Second, persistent laboratories are the most active actors inside the university. Third, the increasing number of contractual agreements signed by persistent laboratories underlines the existence of an ongoing management learning process. Finally, there is no clear-cut difference in the type of contracts signed by persistent and non-persistent laboratories indicating that the development of university-industry relationships is demand-driven. These results are framed within the debate on the ongoing changes in the role and behaviour of universities.

Key words: University-industry relationships; Technology transfer; University research

1. INTRODUCTION

A general reconsideration of the rationale for funding and governance of science is currently underway in both developed and developing countries

F. Cesaroni et al. (eds.), R&D, Innovation and Competitiveness in the European Chemical Industry, 145-173.

(Geuna, Salter and Steinmueller, 2003). Universities across the world are rethinking their roles and activities. In the early 1980s, as a result of government pressure and new technological opportunities, universities embarked on a process of redefinition of their role in society. The traditional view of the university as the locus of scholarly work, research and teaching is being challenged by a more proactive view that sees the university as a central actor in the process of innovation responsible for, and expected to achieve, technology transfer to industry.

A large body of policy and academic literature heralded the arrival of the 'entrepreneurial university', which, via the development of 'third stream activities' or its 'third mission', will become the 'new engine of growth' (Etzkowitz and Leydesdorff, 1997; OECD, 2002; Molas-Gallart et al., 2002). Some scholars have developed concepts such as Mode 2 learning (Gibbons et al., 1994) and the Triple Helix (Leydesdorff and Etzkowitz, 1996) in an attempt to provide a normative framework for the new activities being undertaken by universities. While traditionally universities were expected to contribute to the development of industrial technology via the production of knowledge that was made available in the public domain, the new entrepreneurial university must contribute to (and is explicitly evaluated in terms of) industrial innovation through technology transfer activities such as patenting, creation of spin-off, development of technologies, etc. To achieve this, a series of incentive mechanisms that attempt to align the incentives of researchers with industrial needs need to be developed.

However, there is no agreement about the value of these new activities. Few scholars have developed critical views on the new trends. Some have pointed out that the changes in university behaviour associated with the changes in the incentive structure could have long-term negative unintended consequences (Dasgupta and David, 1994; Geuna, 1999, 2001; Berens and Gray, 2001). Others argue that the over-emphasis on commercialisation could reduce the flow of knowledge from universities and that universities are better producers of talent (via education and graduate research) than of technology (Florida, 1999; Salter et al., 2000; Pavitt, 2001). Others dispute that university patenting is a tool of efficient technology transfer and point out that for most universities these activities will produce net losses (Nelson, 2001; Mowery et al. 2001; Geuna and Nesta, 2003).

There is very little empirical research assessing and evaluating in a systematic way the importance and consequences of the new activities carried out by universities. In Europe very few statistical studies have been produced. To the best of our knowledge, only three studies, one by Gulbrandsen and Smeby (2002) on Norwegian universities, one by Ranga (2003) on the Katholik

Universiteit van Leuven (KUL) in Belgium and one by Caro, de Lucio and Gracia (2003) on the Polytechnic University of Valencia, based on statistical data, have examined the impact on academic research of increased reliance on industrial funding.

The current study aims to contribute to the debate by examining the evolution of contractual relationships at the University Louis Pasteur (ULP) of Strasbourg during the 1990s. We analyse the contracts (including research contracts, cooperative research, services and grants) carried out by researchers employed by university laboratories. Specifically, we focus on the contracts carried out by university laboratories in the chemistry domain. This domain broadly covers contracts in chemistry, physics of materials and pharmacy.

We used an original database that includes 1,177 contracts signed by ULP during the nine year period 1990-1998. After looking at overall trends, we characterise the contracts in terms of their duration, value and typology and examine the evolution of these characteristics during three sub-periods. Our objective is not to assess the consequences of university-industry relationships on the overall activities of the university, but rather, to measure the importance and the evolution of the phenomenon at the laboratory level. We put forward the concept of persistent collaboration (a laboratory with several repeated contracts throughout the period is considered to be a persistent collaborator) as a focusing device to characterise the differences in contracting behaviours. We intend to show that 'third stream activities' have not been developed uniformly inside the university: some laboratories have continuously and increasingly signed contracts with industry over the decade and developed specific competences in contract management, the behaviour of others has been more sporadic. While the importance of the phenomenon is heterogeneous across labs, they all involve a wide range of activities answering a diversified demand. In other words, the supply of university laboratories is driven more by demand than by their own experience in establishing contracts with external partners, even though both influences co-exist.

The chapter is organised as follows. Section 2 briefly reviews the recent literature on university-industry relations. Section 3 provides an overall description of ULP Strasbourg and its contractual relationships. The descriptive analysis of the characteristics of the contracts carried out in the chemistry domain is developed in Section 4. Section 5 adopts an industrial economics approach to the analysis of entry and exit in the '"market"' or game of university-industry collaboration. In so doing we test for alternative interpretations of the changes that are occurring in university contract research behaviour. Finally, Section 6 presents the conclusions from our analysis.

2. UNIVERSITY-INDUSTRY RELATIONSHIPS: THE CORE TOPICS

There is a huge and still growing literature focusing on university-industry relationships. The objective of this section is to show how our contribution relates to this literature. We do not pretend that our scheme is exhaustive, but we think that an appropriate way to present the different approaches is to focus on the three elements that characterise university-industry relationships. The first element concerns firms. Several studies have looked at the economic impact of scientific research on the private sector; the importance of the university as an external source of information used by firms to innovate; and the characteristics of firms that benefit from academic research. The second approach is to study the channels used by both actors - university and industry - to exchange information, and the conditions favouring or blocking these interactions. The third approach focuses on universities and analyses the influence of these alliances on their evolution.

2.1 Firms' point of view

During the '90s, a large number of studies looked at the impact of scientific research results on the competitiveness of firms, industries or nations. These studies are mainly based on surveys, patents and publications. Mansfield (1991) conducted a survey amongst firms in the US that showed that 11% of new products and 9% of new processes could never have been developed or would have come much later in the absence of academic research. Beise and Stahl (1999) replicated this survey in Germany and obtained very similar results. These positive impacts of university research are further confirmed by studies based on patents and publications. The geographical dimension of spillovers is developed in this literature particularly in Jaffe's (1989) work.[23] Most of the studies in question show that geographical proximity matters and that the benefits from academic research increase when the distance between industry and university is small. For instance, Mansfield and Lee (1996) found that firms located near to academic research centres are at an advantage over those located at some distance. Narin et al. (1997) showed that the knowledge flow between the two worlds increased threefold in the US between 1987 and 1994. However, although geographical spillovers seem to be of importance in the US, the results are less clear-cut for Europe: Beise & Stahl (1999) refute the hypothesis and

[23] In this connection, Feldman (1999), Salter and Martin (2001) and Breschi and Lissoni (2001) are valuable surveys.

Arundel and Geuna (2004) claim that localisation matters at country level due to cultural and linguistic specificities in the EU countries.

Several studies (Cohen et al. 2002, Fontana, Geuna and Matt 2004, the Community Innovation Surveys (CIS) I, II and III) examine a broad range of information sources used by firms to innovate, of which one is the research and development (R&D) conducted in Public Research Organisations (PROs). Some distinguish between the sources of information that contribute to innovative ideas and to the completion of innovation. Overall, PROs are not considered to be a major source of information for the innovative process of firms. They do not play a central role in suggesting new ideas. They seem to be more important for innovation completion than for new ideas, but, in both phases, public research is less important than contributions from the vertical chain of production (suppliers, buyers, the firm itself).

Based on surveys, some contributions analyse the characteristics of firms exhibiting a higher propensity to draw upon public research. Cohen et al. (2002) take size and age of the firm as the two explanatory variables. Other studies (Arundel and Geuna, 2004, Shartinger et al., 2001, Laursen and Salter, 2003, Fontana, Geuna and Matt, 2004) incorporate additional explanatory variables, such as the level of R&D expenditure, the degree of firms' innovativeness, the search strategies adopted by firms and their degree of openness. Drawing on CIS II, Mohnen and Hoareau (2003) show that firms cooperating with universities are generally large, are active in scientific sectors, patent and receive government support. Fontana, Geuna and Matt (2004) also highlight factors (public subsidy, patents and R&D level) influencing the number of relationships firms establish with PROs.

2.2 Channels used by both actors to interact

Analysis of the channels used by both actors to exchange knowledge constitutes a central research topic: that is, to understand how the partners interact.[24] For instance, Meyer-Krahmer and Schmoch (1998) asked academic researchers to assess the importance of different types of relationships: collaborative research and informal contacts obtained the highest scores. Cohen et al. (2002) find that the channels of open science, especially publications, public meeting and conferences, but also informal information exchange and consulting, are very important. The results of the Policies, Appropriability and

[24] For a theoretical discussion about the different ways used by universities and firms to form partnerships and their implications in terms of market failure mitigation see Poyago-Theotoky et al. (2002).

Competitiveness for European Enterprises (PACE) survey highlight that large companies see recruitment of new graduates, informal contacts and contract research as the most important ways to access academic knowledge, and that 'low tech' sectors favour formal collaboration much more than do 'high tech' ones (Arundel and Geuna, 2004). The unweighted results of the PACE survey show that publications are the most important method used to draw on public research. The CIS I and II surveys reveal that firms consider universities as important partners in R&D agreements.

Recently, an EC benchmarking study analysed the conditions that facilitate or block high levels of interaction. Lower levels of university-industry relationships can be attributed to lack of demand on the industry side and also to the lack of incentive structures and institutional factors on the science side. High levels of interaction occur when industry demand is high; incentive schemes in science are well developed; the legislation is favourable; and public initiatives to foster university-industry relations are developed along a long-term perspective. University-industry relations differ largely by fields of technology, types of science organisations and enterprises.

2.3 University point of view

Poyago-Theotoky et al. (2002) synthesise the potential benefits and drawbacks of university-industry partnerships for universities. Some benefits are linked to revenue sources such as licensing, equity revenue, sponsored research, donations, etc. Universities may also benefit from the technologies developed by firms (Brooks, 1994, Siegel et al., 2002). Stephan (2001) underlines that "faculty ties with industry have potential for updating the curriculum and initiating new programs" (p. 199). University-industry links may impact positively on students through job opportunities. However, Stephan also points out that these alliances may have negative effects on education (content and quality of teaching), students and curriculum. Technology transfer may interfere with the relationship of trust between faculty and doctoral or post-doctoral students, mainly because of the monetary rewards involved in some inventions. University-industry links may also generate negative effects (Nelson 2001) on the culture of "open science" (Dasgupta and David, 1994). Increased commercialisation of university research can delay the publication of scientific findings and reduce the willingness of faculty to disclose their research agenda and results (Blumenthal et al., 1997). Technology transfer could encourage a shift from basic to more applied research. There is some empirical evidence that rejects this hypothesis and instead shows that entrepreneurial scientists achieve higher productivity scores than their non-entrepreneurial colleagues (Louis et al., 2001).

Some recent theoretical investigations have analysed the incentives for universities to sign specific contracts with firms. For instance, Beath et al. (2000) examine the incentives for universities to get involved in consultancy agreements with private companies, while Beath et al. (2002) study the incentives for both the firm and the university to enter into a research joint venture. These theoretical models are an important contribution to the understanding of the behaviour of universities (their incentives, characteristics, etc.) entering into relationships with firms. However, the role of incentives has been developed more in relation to firms than to universities.

2.4 Our contribution

Our study refers mainly to the last two approaches: we analyse different types of communication channels (grants, research contracts, research collaboration, services and other special contracts) used by laboratories to establish links with firms and other institutions over a decade (1990-1998). We base our analysis on a portfolio of contracts involving 52 ULP laboratories in the chemistry domain.

Our purpose is not to analyse the potential positive or negative effects of university-industry relationships on the university, but to characterise in an empirical way the importance of the phenomenon at the laboratory level. University-industry relationships have always constituted a part of the activities carried out by some laboratories; government pressure to increase the interaction between universities and firms during the '90s induced more laboratories to carry out research for firms. Therefore, we could expect heterogeneous behaviour at the level of the laboratory dependent on the history of the involvement of the laboratory in contractual research. Thus, the first element of originality is not to consider the university as a whole, but to underline the differences in the contracting behaviours of a set of chemical laboratories and to identify the existence (or not) of core laboratories in signing contracts (i.e. persistent laboratories). The second original contribution relates to the empirical exploration of whether the historical involvement of laboratories with firms has affected the characteristics of the portfolio of contracts established by these laboratories with external partners. Our intuition is that greater experience in contracting (i.e. persistency) induces a specialisation in longer-term research contracts with relatively high budgets rather than shorter-term, service oriented activities. On the other hand, if the portfolios of persistent and non-persistent laboratories are similar and diversified, we would suggest that the demand probably heavily influences the type of contractual activity carried out by the laboratories. In this second case we would be witnessing a demand-led process of development of university-industry relationships. Being able to distinguish between the two is important from a policy perspective because it will improve the efficacy of policy actions.

From a methodological point of view, we use entry, exit and persistence indicators to highlight differences in the contracting behaviour of laboratories over the nine year period and to underline that persistence in contracting is a major phenomenon. We use appropriate statistical tests to analyse whether persistent and non-persistent laboratories have the same portfolio of contracts and whether the demand side or the experience hypothesis dominates, or both co-exist. Overall, our contribution aims at shedding new light on how past behaviour in terms of contracting influences the current portfolio of contracts of academic laboratories.

3. ULP CONTRACTUAL RELATIONSHIPS

This section examines the importance of and evolution of ULP research laboratories' contractual relationships. We mainly focus on non-EU contracts in the broadly defined chemistry domain. By non-EU contracts, we mean contracts negotiated outside any of the European Union framework programmes, i.e. all contracts signed between a ULP laboratory and an external private (e.g. a firm) or public institution.

3.1 ULP - A general presentation[25]

ULP, located in Strasbourg, France, is historically a research university. The German Emperor, Kaiser Wilhelm, gave the first major impulse to its development when the Alsace region was annexed to Germany, between 1870 and 1918. The university was restructured, both in terms of infrastructure and research direction and organisation, following the 'Humboldt' principle focusing on natural and experimental sciences. The investments and general focus were maintained when the university reverted to French administration, with the addition of life sciences, which was developed mainly after the 2nd World War. The re-organisation of French universities after 1968 brought about the constitution of three independent universities, ULP being one of them. The

[25] This section draws on Llerena and Meyer-Krahmer (2003).

core departments (or similar organisational structures) of ULP are: mathematics and computer sciences; natural sciences (chemistry and physics); life sciences (medicine, pharmacy, ontology, neuroscience and biology); social sciences (psychology, educational sciences and economics and management); engineering sciences with four engineering schools in biotechnology, physics, earth sciences and chemistry and a technical institute.

ULP is, in the French context, very specific not only because of its history but also because of its structure. Since the 19th century, its main emphasis has been research excellence (especially of academic research) and international openness. Although ULP is a medium sized university in terms of number of students (17,000 in 2003), it has the largest consolidated budget of the French universities (almost 200 million Euros, consolidated budget, i.e. including the salaries of public servants). Only 60% of the budget originates from direct government subsidies (the national average being 80% to 95%), and the budget devoted to research represents around 35% (logistics excluded). These structural characteristics are explained for the most part by the important activity of medical and environmental analysis conducted at ULP. ULP has a major responsibility to carry out medical analyses for Strasbourg University Hospital and is responsible (to a lesser degree in terms of budget) for environmental pollution analysis (food and water essentially). The link with the hospital is a heritage from the period of German governance of the university, and is unique in France. It represents about 45 million Euros.

ULP is also generally seen to be internationally oriented compared to other French universities. Twenty-seven per cent of the students are post-graduates students, about 30% of which are foreigners. Around 3,500 staff are engaged in research activities at ULP, including all worker status and all organisational affiliations (such as CNRS and INSERM). Finally, not only is the budget devoted to research important but also more than half is external non-grant based (such as specific contractual arrangements with firms or other institutions, or public tenders and procurement). Thus, ULP is one of the most active universities in the French system in terms of contractual activity.

3.2 The evolution of contractual activities at University Louis Pasteur

Even if we ignore the historical perspective in this analysis, it should be noted that ULP (or more precisely at that period, University of Strasbourg and the Faculties of Sciences) was recognised as having a great number of contractual arrangements between the two World Wars, particularly with oil companies and the agro-food and chemical industries. Our analysis focuses on the contemporary period and covers almost one decade (1990-1998) of contractual activities. The data set was developed by *ULP-Industrie*, the university's Technology Liaison Office (TLO). This office, one of the first of its kind in a French university, was set up at the beginning of the 1980s, which gives us confidence as to the exhaustiveness of the contracts registered in the 1990s (at least for the disciplines under consideration): the internal regulation of the university requires that all contracts must be scrutinised and registered by the TLO before signature by the rector of the university. The database includes 1,177 contracts signed during the nine year period.

Table 1 - Number of Contracts - 1990-1998

Year	Total	EU	Non EU	Annual Growth Rate	CAGR* (year 1990)
1990	37	4	33	0%	
1991	58	4	54	57%	
1992	123	6	117	112%	
1993	120	33	87	-2%	
1994	139	25	114	16%	
1995	170	14	156	22%	
1996	216	52	164	27%	
1997	143	14	129	-34%	
1998	171	11	160	20%	
Total	1,177	163	1014	24%	19%

**Compound annual growth rate*

Table 1 shows the aggregate number of contractual relationships over the period 1990-1998. The number of contracts reaches a peak in 1996; after which there is a slight decrease. This peak is attributable to the entrance and the large weight of EU contracts. Taking into account EU contracts only, there is another peak in 1993, which, when measured against total contracts, is counterbalanced by a decrease in non-EU contracts. Over time, the contractual relationships of ULP show a compound annual growth rate of 19%.

Do EU contracts represent a significant proportion of total contracts between 1990 and 1998? The ratio 'number of EU contract/total number of contracts' varies between a minimum of 6.5% (in 1998) and a maximum of 24% (in 1996), which produces an average of 13% over the whole period. We can say, therefore, that EU contracts make a significant contribution to the total. Nevertheless, within this category there are different typologies of contracts, such as grants, research contracts and networks, and different configurations of actors. The database contains information concerning the kind of contracts, but not information about the different partners involved. As the unit of analysis is the firm and the ULP laboratory, we had to exclude from the analysis the EU contracts, despite their importance. Yet, the large weight of EU contracts is an additional indicator that ULP is an internationally oriented research university.

What are the main active research domains in terms of contractual relationships? Where possible, we classified contracts according to five main research domains: Chemistry, Life and Earth Sciences, Mathematics and Computer Sciences, Human and Social Sciences, and Others. The broadly defined chemistry domain includes contracts in chemistry, physics of materials and pharmacy, along with all the contracts where the research domain was not specified, but in which the research subject (a more disaggregated classification) was clearly within the chemistry domain. What emerges is the overwhelming leadership of the chemistry research domain. Moreover, the relative weight of experimental sciences compared to social sciences is clearly evident. On average, over the whole period, the number of contracts in chemistry accounts for almost 57% of the total (see Table 2). The proportion of EU contracts in chemistry is approximately equal to the proportion of EU contracts in the total number of contracts signed in all the scientific fields. In other words, chemistry does not benefit from EU support any more than the other domains.

Table 2. Share of chemistry domain and share of EU contracts in chemistry

Year	Chemistry/total ULP	EU Chemistry/total Chemistry
1990	86.5	3.1
1991	48.3	0
1992	58.5	1.4
1993	50.8	42.6
1994	43.9	21.8
1995	54.1	9.8
1996	59.3	23.4
1997	61.5	7.9
1998	59.6	5.9
Total	56.8	13.2

We have shown that chemistry is the largest domain within ULP in terms of number of contracts (and monetary value) and that it has similar characteristics in terms of EU contracts as compared to the other domains examined. As we lack the specific information for the EU contracts that is needed for the analysis at the level of ULP laboratories and external private (i.e. firm) and public organisations, we focused on non-EU contracts in the chemistry domain.

4. CHARACTERISTICS OF CHEMISTRY DOMAIN CONTRACTS

The contracts between ULP laboratories and firms or other institutions have been analysed in terms of their duration, nature of the relationship and monetary value.[26] We developed an analysis of the changes in these characteristics across the three sub-periods: 1990-1992; 1993-1995; 1996-1998. This allowed us to make a first approximation of the evolution of the contractual relationships of ULP laboratories with their partners.

Over the whole period, 52 laboratories in the chemistry domain signed 571 contracts with 247 firms and other institutions (i.e. an average of 11 contracts per laboratory). Table 3 shows the number of laboratories involved in contracts in each sub period, the number of contracts signed and the concentration

[26] Here the term 'contract' means a contractual relationship between the university and firms or other institutions, which might take the form of a research contract, research collaboration, grant, provision of services or some other type of contract not included in the above categories. In the text we use the term contracts to embrace all these categories, unless otherwise indicated in the analysis of the particular nature of the contractual relationship.

indexes C4 and C10 i.e. the concentration of chemistry contracts by laboratory.[27]

Table 3 - Chemistry contracts by laboratories – Three sub-periods

	1990-92	1993-95	1996-98
No of ULP labs	25	31 (+24%)	43 (+39%)
No of contracts	130	166 (+28%)	275 (+66%)
C4 (contracts)	0.49	0.52	0.44
C10	0.75	0.73	0.64

The concentration of chemistry contracts is similar in the first two sub-periods. Over these two periods, the number of laboratories increased roughly proportionally to the number of contracts. C4 shows that the first four laboratories accounted for roughly half of the total number of contracts. The 10 most important laboratories in terms of relationships accounted for approximately 75% of all the contracts in the two periods. It should be noted, however, that the top ten and the top four laboratories vary across the two sub periods.

The last sub period is characterised by an increase in terms of the number of laboratories involved in contractual relationships and also, more importantly, by an increase in the number of contracts signed. These changes result in a decrease of contracts concentration for the top four and the top ten laboratories: C4 and C10 are respectively 0.44 and 0.64. The distribution of contracts across laboratories is less asymmetric than in the first two sub periods.

4.1 How does the duration of links between ULP laboratories and their partners evolve?

Figure 1 highlights a polarisation. Over the three sub-periods, the number of the shortest (less than 6 months) and the longest (more than 37 months) contracts increases while the number of contracts that last between 7 and 12 months decreases. This would indicate a preference for very short or very long contracts, which might correspond to different kinds of activities. The interpretation of these evolutions is difficult and would imply further research. It seems that laboratories under financial pressure have to accept short-term contracts, but try to be selective and develop long-term partnerships.

[27] We also calculated the Herfindhal index for each period: 1990-92 (H=0.08 ; 1/N=0.04) ; 1993-95 (H=0.11 ; 1/N=0.03) ; 1996-98 (H=0.07 ; 1/N=0.02). This confirms that the last period is the least concentrated.

Figure 1. Contract length

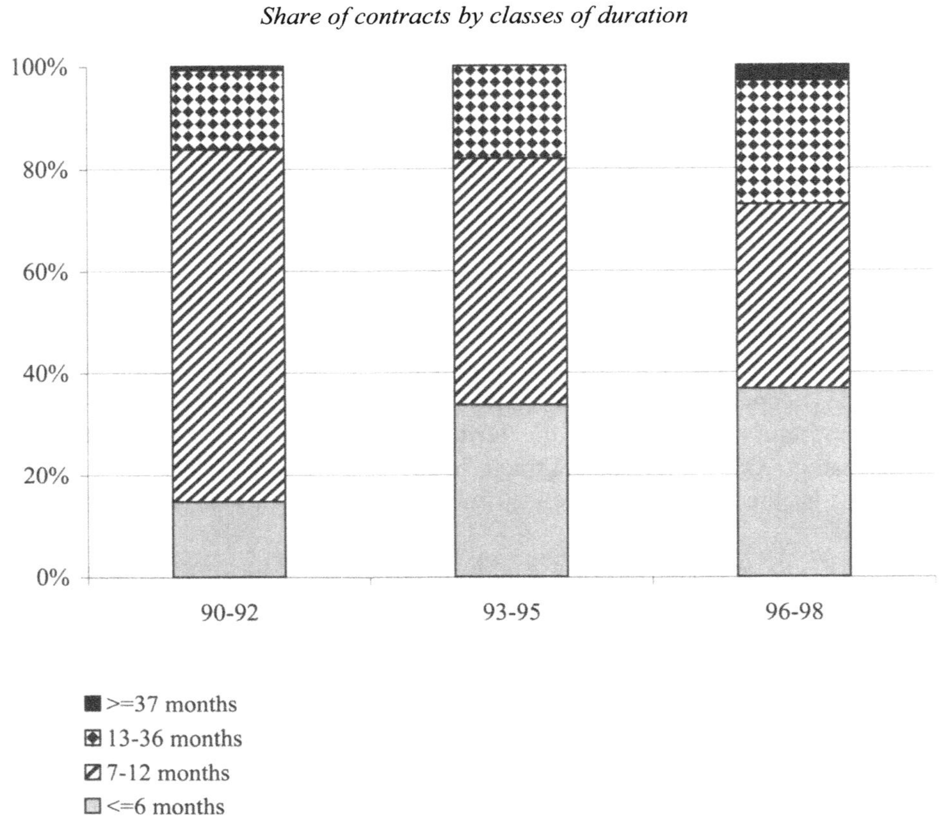

4.2 Does the type of contract with university laboratories change?

Over the three sub-periods, the share of services in total contractual activities increases (see Figure 2). Research activities (research contracts and research collaborations) seem to be stable over the period considered, with an increase in the proportion of research collaborations in the third sub-period. The share of grants (essentially for doctoral and post-doctoral researchers) shows an increase in the first sub-period and a substantial decrease in the third.

Figure 2. Number and type of contracts

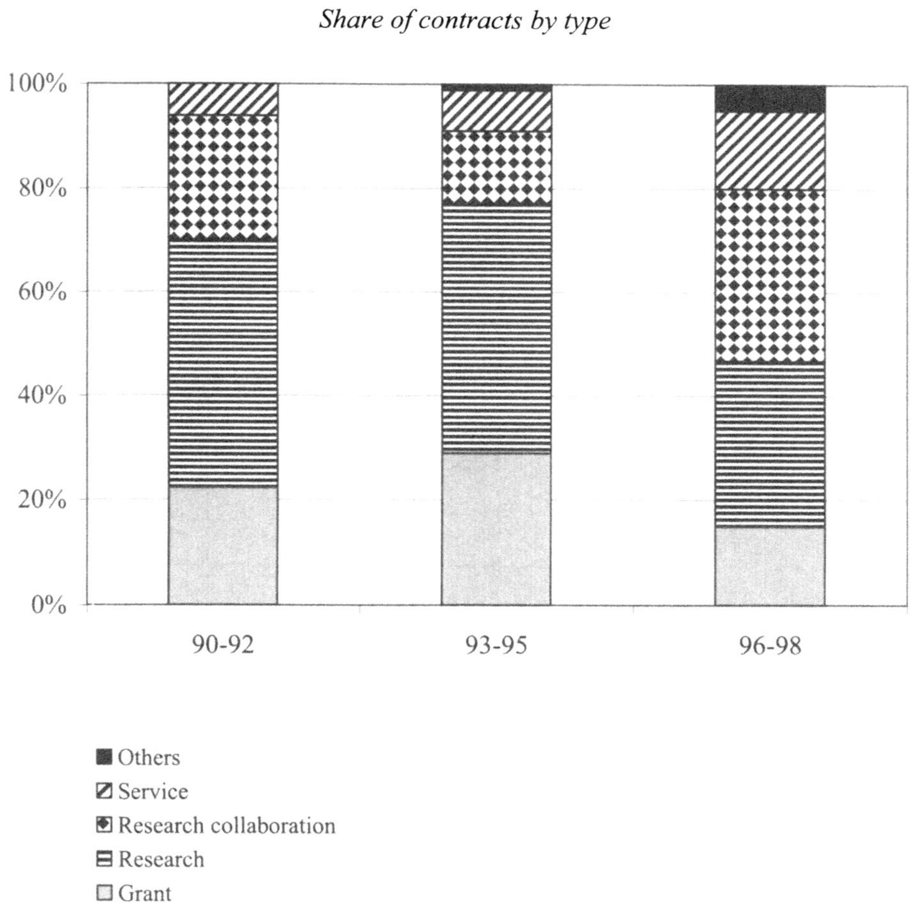

Figure 3 shows the changes in types of contracts in terms of value. The polarisation of types of contract over the nine years period is even more evident when the value of the different types of contracts is taken into account.

Figure 3. Type of contracts (in Euro CPI)

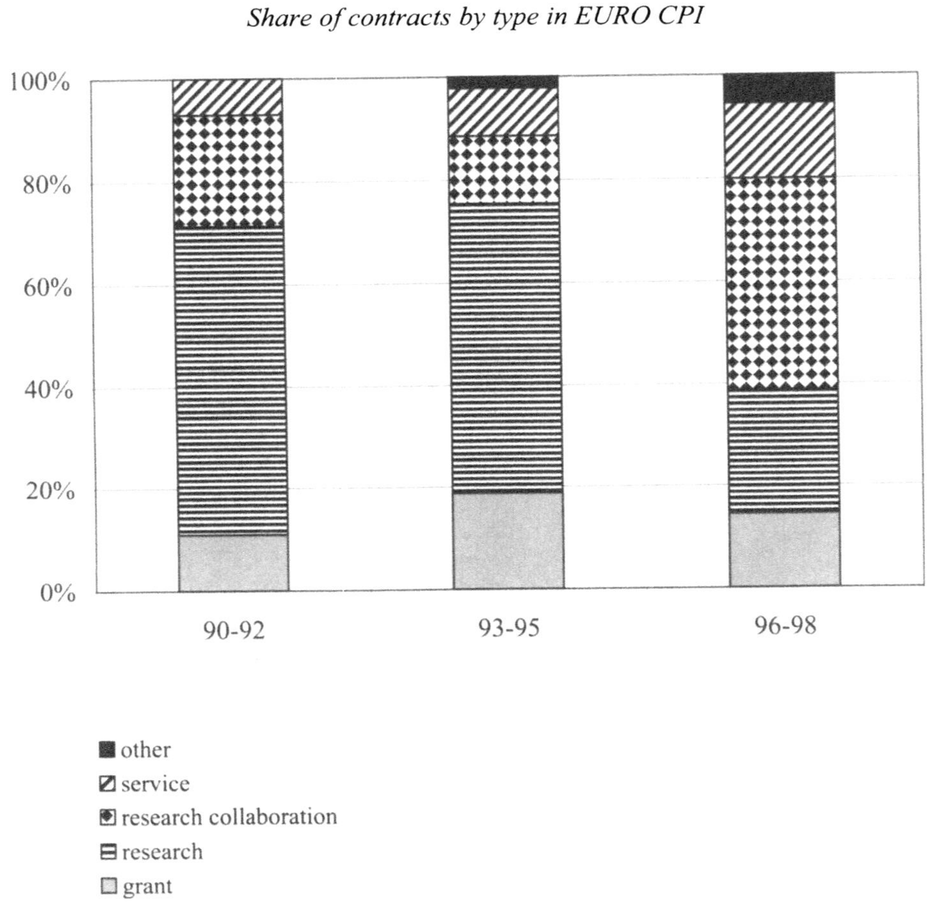

4.3 What are the changes in monetary value of contracts?

We have classified contracts into six classes according to their budget from small to large: A to F. Classes A and B include the low-budget contracts, C and D medium-sized ones and E and F the high-budget contracts. The classes are calculated on the basis of 1990 constant prices. Figure 4 shows the same polarisation phenomenon that emerged for contract length (cf. Figure 1). Over the three sub-periods, the smallest contracts (especially class A[28]) and the largest contracts increase in number, while the number of medium sized contracts decreases.

In summary, the number of contractual relationships increases over the nine year period both in terms of number of ULP laboratories involved and contracts signed; this produces a decrease in concentration of chemistry contracts by laboratory. A double polarisation phenomenon occurs for contract length and budget (medium contract length and budget decrease in favour of the extreme classes). The share of services increases and research activities seem to remain stable.

Figure 4. Contract budgets

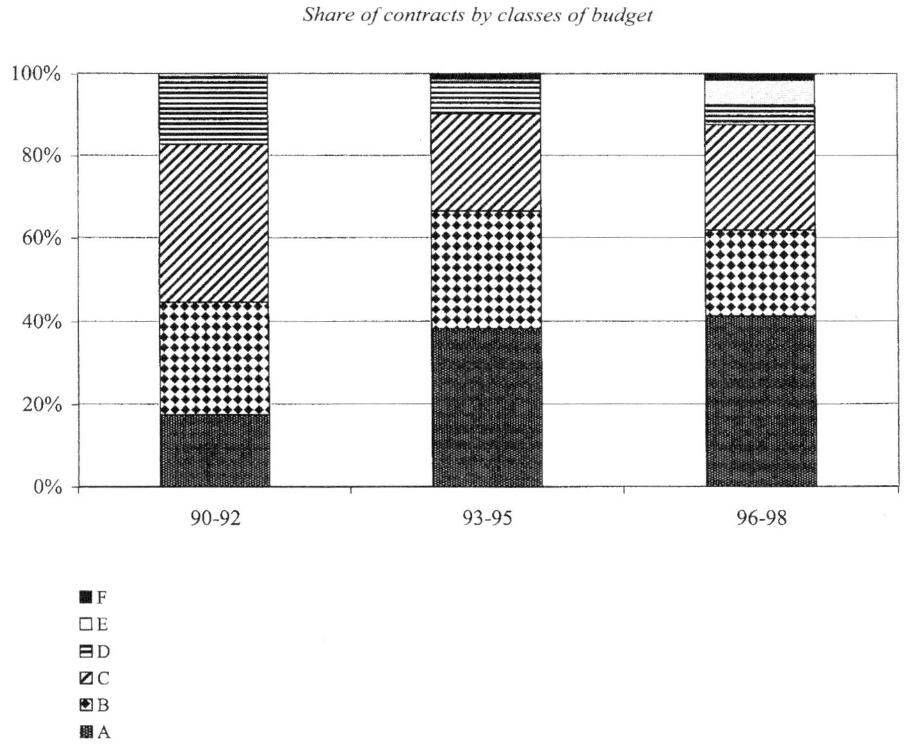

[28] Note that in class A, most of the contracts are contracts with value 'zero', because they correspond essentially to service contracts, implementing an agreement on the unit costs of the service (for example, a particular analysis), the number of units executed being defined in real time during the year depending on the needs of the client.

5. ENTRY, EXIT AND PERSISTENCE OF COLLABORATION

During the period under examination the participation of ULP laboratories in contracts with firms and other institutions (F&I) increased, and the duration, nature of the relationship and monetary value of the contract also changed significantly. These changes in the characteristics of the contracts could be attributable to two different phenomena. First, the entry of new laboratories in the contractual relationship with F&I 'market' might have changed the average characteristics of the total number of contracts carried out. Collaborations between similar organisations in terms of objectives and activities (among firms) and, even more so, collaborations between different organisations (firms and universities) are characterised by learning processes and confidence/trust building. New laboratories, hitherto unused to having contracts with F&I, started to develop such collaborations to counter the shortage of public support and/or in response to the opportunities offered by F&I. It might be hypothesised that, being inexperienced, they began by taking on occasional contracts of small size, with a short time horizon and more service oriented. At the other extreme, laboratories with a longer experience of collaboration with F&I could be expected to have continuous (more persistent), larger, and longer-term contracts more focused on research.

An alternative explanatory hypothesis is that the changes in the average characteristics of contracts might be due to a structural change in the demand for collaboration by F&I with universities. In Section 4 we showed that all types of laboratories, those 'new to the game' and those with experience and more persistent collaborations, shifted over time towards smaller sized, shorter duration and more service-oriented types of contracts. This is a sign of structural change in the relationship between universities and firms.

To examine these two alternatives we deploy an industrial economics approach to the analysis of entry and exit in the market or game of university-industry collaboration, focusing first on the F&I and then on the ULP laboratories. As in the previous analysis, we consider the three sub-periods: 1990-1992; 1993-1995; 1996-1998. The analysis aims at exploring whether some laboratories have more continuous collaboration with F&I, that is, whether they collaborate in a persistent way with F&I. On the basis of the results of this analysis we will be able to provide some evidence in support of the two alternative interpretations of the changes in the characteristics of the contracts proposed above. If the first hypothesis is true, the characteristics of the contracts signed by those laboratories and F&I that have had a persistent relationship during the period should be different from the characteristics of those contracts signed by laboratories and F&I that were new to university-industry collaborations. If the second hypothesis is true, no significant difference should be found in the characteristics of the contracts signed based on the history of involvement in contractual relationships.

The two alternative interpretative hypotheses on the behaviour of laboratories over time ((i) laboratories successfully developed the learning necessary for the management of their contractual relationships, which allowed them to specialise in larger and longer-term contracts more focused on research; (ii) they simply reacted to changes in demand regardless of their historical involvement in contractual relationships) have different policy implications. Indeed, most current policies focus on creating incentives for universities to increase their interactions, i.e. policies are aimed at stimulating the responsiveness of public laboratories to the needs of firms and other organisations. If universities are already responsive to changes in demand and, across laboratories, are carrying out shorter-term service oriented contracts (our second hypothesis) the policy aims would already have been realised and policies aiming at increasing the comparative advantage of universities, shaping the university supply towards long-term research contracts, would be needed.

We start by briefly analysing persistent F&I; we then move on to the detailed examination of persistent laboratories followed by examination of the interaction between persistent firms and persistent laboratories. We conclude this section with a first attempt to test the validity of the two alternative interpretations suggested above.

5.1 Entry, exit and persistence in collaborations: F&I

We identify these three categories of F&I:

- an entrant is an F&I with no contracts with a university laboratory during the previous sub-period. Thus, an entrant in period 2, is an F&I with no contracts in period 1 and an entrant in period 3 is one with no contracts in period 2.
- a stable or incumbent F&I, is an organisation that has worked under contract with a university laboratory during the previous period. An incumbent in period 3 corresponds to an F&I that has collaborated under contract with a university laboratory during period 2, but not necessarily during period 1.
- a persistent F&I is an organisation with contracts over the three periods.

Table 4. Stability of actors' presence - F&I (no of contracts)

	1990-92	1993-95	1996-98
Total no present	38 (130)	89 (167)	133 (275)
No of entries		74 (108)	102 (149)
Stable presence		15 (59)	31 (126)
Stable presence firms only		12 (49)	24 (69)
Persistent presence	11 (60)	11 (53)	11 (88)
Persistent presence firms only	8 (44)	8 (43)	8 (63)

Stable presence: F&I continuously present in two sub-periods

Persistent presence: F&I continuously present in three sub-periods

Table 4 presents the evolution of these three categories and the number of contracts they have been involved in. In the first sub-period 38 F&I entered into contractual partnerships with university laboratories and, thus, were present in the first three years. Among the 89 F&I with contracts in the second sub-period, 14 were also present in the first, one of which signed a contract of long duration during that first sub period. There are thus 15 F&I that could be considered as having a stable relationship in the first two sub-periods. During the last three years, 133 F&I signed new contracts and 31 F&I had had contracts in the second sub-period so these 31 can be considered to be incumbent or stable organisations.

The number of F&I, which continuously signed contracts all over the three sub-periods is11 - eight firms and three other institutions. We define these 11 as having a persistent relationship. In the first period they accounted for about 68% of all contracts, while in the second and third periods they accounted for only 32%. Overall, we can see a significant entry of new F&I, involved in about 1.4 contracts in each sub-period. Stable F&I tend to be much more active than new entry with about four contracts in each sub-period. Finally, persistent F&I are those institutions with more contracts - ranging from a minimum of about five each to a maximum of about eight - in the three sub-periods. Specifically, persistent firms are those organisations with more contracts.

5.2 Entry, exit and persistence in collaborations: university laboratories

To analyse the entry, exit and persistence of ULP laboratories in university-industry collaborations we focus only on university-firm contractual links, excluding other institutions, i.e. national and foreign public research institutions

(Centre National de la Recherche Scientifique –(CNRS), Centre d'Energie Atomique (CEA), etc.), universities, ministries, regional councils, the EC, non-profit organisations, etc. Our choice is motivated by three issues: (i) the population of other institutions is heterogeneous in terms of types of organisations; (ii) we are primarily interested in university-industry links that are the most widely studied in the literature; (iii) in the previous analysis we have shown that among F&I firms are the more persistent and account for more contracts. As this chapter centres on the concept of persistent collaboration as a focusing device, we concentrate on the persistent organisations.

This restriction results in the loss of six laboratories out of 52 and 130 contracts out of 572. This means that over the decade six laboratories entered into contracts only with other institutions. As in the previous analysis three categories are identified:

- an entrant is a university laboratory with no contracts during the previous sub-period.
- a stable or incumbent laboratory is one with a contract with industry during the previous period.
- a persistent laboratory is one with contracts in each of the three periods.

Table 5. Stability of actors' presence - laboratories (no of contracts)

	1990-92	1993-95	1996-98
Total no present	22 (107)	29 (139)	35 (196)
No of entries		8 (21)	12 (31)
Stable presence		21 (118)	23 (165)
Persistent presence	18 (84)	18 (115)	18 (150)

Stable Presence: Laboratories continuously present in two sub-periods
Persistent Presence: Laboratories continuously present in three sub-periods

Table 5 shows the evolution of three categories of ULP laboratories according to their entry-exit behaviour over the three sub-periods. Table 5 highlights the number of entry, incumbent and persistent laboratories, and also the number of contracts they signed. In the first sub-period, 22 ULP laboratories signed 107 contracts with industrial organisations. Out of these 22 laboratories, 21 were incumbents in sub-period 2 as they continued their industrial contracting activities (with 118 contracts, which account for 85% of the contracts in period 2). During the same period, eight laboratories established first time relationships with firms (8 entrants with 21 contracts, which means 15% of all contracts). Thus, in the second sub-period, 29 laboratories (21 stable and 8 entrants, respectively 72% and 28% of the total number of contracts signed) were active with 139 contracts. Out of these 29 actors, 23 (with 165 contracts, which account for 84% of the contracts in sub-period 3) were stable and six had no contractual activities in period 3. In this same sub-period, 12 new actors (with 31 contracts, 16% of all contracts) entered into contractual arrangements. The total number of actors present in the third sub-period is 35 accounting for 196 contracts. In sub-period 3, new entrants represent 34% of the population and stable actors account for the remaining 66%.

In terms of number and share of laboratories, entry is an increasing phenomenon between sub-periods 2 and 3, but entrants signed only 15% of the contracts in each period with an average of 2.6 contracts per laboratory. This could be explained by the fact that there is no learning effect from one period to another and that, for a first experience, university laboratories only accept a relatively small number of research contracts (2.6). Incumbents are the most active in terms of number of contracts in sub periods 2 and 3: they signed 85% of the contracts. The question that arises concerns the persistence of these actors. Are they only stable between two periods or are they also present over the whole nine years?

Is persistence a significant phenomenon? Table 5 shows that 18 laboratories are persistent over the whole period and show an increase in the number of contracts signed. The share of persistent laboratories in the total number shows a decrease (from 82% in period 1 to 51% in period 3); this is due to an increase in the number of laboratories involved in contractual activities. But, more interestingly, the share of signed contracts by persistent laboratories is very important and remains constant over the period (about 80% of contracts). In the last period, 50% of the laboratories are persistent and they are involved in 80% of the contracts. Persistence is obviously a major phenomenon. Persistent laboratories play a much more important role than persistent firms in terms of share of contracts signed.

Persistent laboratories consistently have a higher number of contracts both compared to the total population and in relation to the sub-sample of stable actors. This means that the trend for the total population (an increasing number of contracts) is largely explained by the behaviour of the persistent laboratories. Moreover, for persistent laboratories the number of contracts increases over the period; this probably underlines a learning effect, which is not significant for the other laboratories that only have occasional contracts, confirming the evidence that persistence in term of presence over time is always qualified by persistence in terms of share of contracts signed.

5.3 Persistent relationships between F&I and laboratories

To better qualify the persistence of collaboration between university laboratories and F&I we analyse a map of the 'persistent relationships' that have emerged among the actors so far considered.[29] If we examine the persistent university laboratories on the one side, and the F&I that have signed more than one contract with these laboratories over the three sub-periods considered on the other, we obtain the following results. Overall, the number of persistent laboratories is 18. Of these, 10 have had a persistent-like type of relationship with the sub-sample of persistent F&I, which have signed more than one contract in each sub-period. Moreover, six of the 18 persistent laboratories have co-operated with more than one of these F&I, with one laboratory having contracts with up to eight F&I. We should also stress that in most cases, the role of other institutions is relatively marginal if compared to the persistence of the relationships that university laboratories establish with firms.

5.4 What drives changes in contract characteristics? A preliminary analysis

Given that we are able to identify two distinct groups of laboratories: those with persistent collaborations with firms and those that had only occasional collaborations, we can now turn to a preliminary test of the two alternative interpretations of the reasons behind the changes in contract characteristics. We might reasonably hypothesise that a high degree of persistence leads to mutual learning and development of trust among the different actors involved. This type of relationship gives rise to a structural and stable channel of mutual knowledge flows, usually via long-term, research-based contracts. On the other hand, the instability of contractual relationships might be a symptom of a more dynamic demand-supply trend of services, characterised by short-term contracts of low value and the presence of one-off actors.

Is there any statistical evidence that the laboratories with persistent collaborations have signed contracts with specific characteristics? To test this hypothesis we analyse the contractual characteristics of the 18 laboratories and compare them to the other laboratories. In particular, we compare the two groups of laboratories in terms of time lengths, budgets and types of contracts

[29] Due to a confidentiality agreement we cannot provide the list of names.

signed. We would assume that the persistent laboratories had more time to develop mutual learning and co-specialisation so that they should have contracts with higher budgets, of longer term and less service intensive. We used a t-test and a non-parametric Mann-Whitney to test if there were significant difference in the average length and budget of contracts. There is significant difference in the mean length of the contracts. However, contrary to our expectations, the non-persistent laboratories (those laboratories that had contracts in only one period or in only two periods) on average were involved in longer contracts than those signed by the persistent laboratories (15 months versus 11 months). We did not find any significant difference in the average budget of contracts: persistent and non-persistent laboratories tend to be involved in both small and large budget contracts. The most important difference is in the type of contract. The non-persistent laboratories have a much lower share of grants and a slightly higher share of services.

The results of the tests do not seem to provide a clear-cut answer. On the one hand, there is no clear evidence to support the interpretation that entry into the world of university-industry relationships by new laboratories implies that these laboratories would be involved in shorter term, lower budget service oriented contracts. The non-persistent laboratories have a slightly higher share of services in their portfolio, but they are also the laboratories that are involved in longer-term contracts. On the other hand, there is not sufficient evidence to support the view that the '90s was a period of a structural change in relationships between ULP and firms characterised by changes in the types of contracts with all types of laboratories involved in more demand-supply, service oriented kinds of relationships with firms.

It is important to note that the results of the analysis confirm the existence of relevant learning processes going on in the persistent laboratories. These laboratories signed an increasing number of contracts - by the end of the nine years period they were responsible for about 80% of contracts. Contracts have become a 'routinised' activity, inducing economies of scale due to learning the procedures at the laboratory level; these, probably mainly managerial/organisational economies of scale, are applicable to all types of contracts. Instead, it seems that the new entrants could be characterised as displaying more opportunistic behaviour - they enter into contracts when an interesting subject, and relatively substantial funding, arises - and sometimes, in periods of budgetary constraint, just the funding will be sufficient attraction. This opportunistic behaviour would be consistent with longer-term contracts for the non-persistent laboratories because for them, accepting a long-term contract is more 'similar' to receiving public funding.

6. CONCLUSIONS

This chapter has analysed the evolution of contractual research relationships in the ULP during the 1990s. This research intensive university witnessed an important increase in the number of research contracts carried out for firms and other institutions in the first part of the nineties, followed by a stagnation up to 1999. The most important area of contractual research activity is the chemistry domain (which includes contracts in chemistry, physics of materials and pharmacy).

Comparing the end of the 1990s with the beginning, the type of contract research activity carried out by the university laboratories has changed. The duration of contracts has become more concentrated on either long-term or short-term contracts. Similarly, budgets have become more polarised on either very large or very small budgets, with the latter accounting for 40% of contracts by the end of the nineties (small budget contracts were about 18% in 1990). Finally, the university has carried out a larger number of service type contracts and has been less involved in the delivery of doctoral degrees funded by industry.

Four main conclusions can be drawn from the analysis of the entry and exit of ULP laboratories in the market of contractual activity with F&I. First, a large number of new actors have entered the arena of university-industry relationships. A very large number of firms and other institutions have signed contracts with ULP laboratories for the first time. By the end of the '90s 3.5 times more F&I were collaborating with ULP chemistry laboratories. Similarly, an increasing number of laboratories have been involved in contracts with F&I. Further research is needed to see if, at the laboratory or individual researcher level, the contractual activities represent a substitute or a complement with respect to academic research (measured, for example, by publications). A first tentative assessment (on the basis of a different database not fully comparable for time coverage and reorganisation of the laboratory structure to the one employed here) seems to show that contractual activities are not a substitute for, but rather a complement to current academic research at the laboratory level. However, it is possible that in large laboratories individual researchers undergo a process of specialisation (with substitution effect), making the interpretation of aggregate figures more complicated. Moreover, any assessment of the complementarity between contractual activities and academic research has to take into account their sustainability. Longer time series are needed to include time lags between the contractual activity and the academic research output that would allow the hypothesis that an increased contractual activity does not affect the future production of academic research (for example via a reduced accumulation of human capital for those researchers that carry out contractual activities) to be tested.

Second, the concept of persistence is a useful focusing device to analyse university-industry relationships. In the case of the laboratories, those that signed contracts throughout the nine years period examined, i.e. the persistent ones, are the most active players. By the end of that period they account for about 80% of the contracts and also have multiple contracts in the three sub-periods. Though less marked, in the case of F&I, persistent institutions also play an important role; they are the ones with the highest number of multiple participations in contracts though their total share of contracts decreased during the period due to the entry of a large number of new F&I.

Third, we did not identify a clear-cut difference in the type of contracts carried out by persistent and non-persistent laboratories, the second group tend to be more involved in services and less in grants, but also tend to have longer term contracts. It seems therefore that two phenomena coexist. All laboratories tend to be involved in broader types of contracts with firms and other institutions, as the demand for university services and research has become more diversified. At the same time, new laboratories have started to collaborate with firms by way of delivering services rather than hosting PhD students doing a thesis funded by firms (the traditional way of cooperation in the French context).

Finally we identified some evidence to support the view that an important process of learning is going on in the persistent laboratories. Those laboratories that had contracts all along the considered period seem to have developed specific competences in managing contracts that allowed them to obtain an increasing number of contracts by the end of the period considered. However, these contracts were not significantly different from those of the non-persistent laboratories.

The evidence put forward in this chapter highlights that the University Louis Pasteur of Strasbourg has undergone a period of change in the type of research activities carried out. By the end of the 1990s, university laboratories had signed four times more research contracts than at the beginning of the nineties and they were involved in a much larger share of short-term service oriented type work. Two main typologies of laboratories have been identified, the persistent ones responsible for the largest share of contracts and active all along the period, and the new entry with spot collaborations. We were not able to identify clear cut differences in the type of contracts entered into by the two typologies of laboratories. It seems that both are affected by the overall changes going on within the university. Further research taking into account the publication and patenting output of these laboratories at the researcher level should be developed to better understand the implications of the ongoing changes.

NOTES

[1] Comments and suggestions made by Alfonso Gambardella, Walter Garcia-Fontes, Ed Steinmueller, the participants in the Third Triple Helix Conference in Rio de Janeiro and the participants in the City University (London) seminar series are acknowledged. Particular thanks go to Claudia Rutigliano for her skilful research assistance. The financial support of the Commission of the European Communities, TSER project "From Science to Product" is also acknowledged.

[2] In this connection, Feldman (1999), Salter and Martin (2001) and Breschi and Lissoni (2001) are valuable surveys.

[3] For a theoretical discussion about the different ways used by universities and firms to form partnerships and their implications in terms of market failure mitigation see Poyago-Theotoky et al. (2002).

[4] This section draws on Llerena and Meyer-Krahmer (2003).

[5] Here the term 'contract' means a contractual relationship between the university and firms or other institutions, which might take the form of a research contract, research collaboration, grant, provision of services or some other type of contract not included in the above categories. In the text we use the term contracts to embrace all these categories, unless otherwise indicated in the analysis of the particular nature of the contractual relationship.

[6] We also calculated the Herfindhal index for each period: 1990-92 (H=0.08 ; 1/N=0.04) ; 1993-95 (H=0.11 ; 1/N=0.03) ; 1996-98 (H=0.07 ; 1/N=0.02). This confirms that the last period is the least concentrated.

[7] Note that in class A, most of the contracts are contracts with value 'zero', because they correspond essentially to service contracts, implementing an agreement on the unit costs of the service (for example, a particular analysis), the number of units executed being defined in real time during the year depending on the needs of the client.

[8] Due to a confidentiality agreement we cannot provide the list of names.

7. REFERENCES

Arundel, A. and A. Geuna. 2004. Proximity and the use of public science by innovative European firms. Forthcoming in Economics of Innovation and new Technology.

Beath, J. Owen, R. Poyago-Theotoky, J., and D. Ulph 2000. Optimal incentives for income generation within universities. University of Nottingham, School of economics, Discussion Paper 00/16.

Beath J., Katsoulacos, Y. Poyago-Theotoky, J. and D. Ulph 2002. University-firm cooperation and knowledge transfer. Mimeo.

Behrens, T.R. and D.O.Gray. 2001. Unintended consequences of cooperative research: Impact of industry sponsorship on climate for academic freedom and other graduate student outcome. Research Policy. 30: 179-199.

Beise, M. and H. Stahl. 1999. Public research and industrial innovation in Germany. Research Policy. 28: 397-422.

Blumenthal, D. Campbell, E.G. Anderson, M.S. Causino, N. and K.S. Louis. 1997. Withholding research results in academic life science: evidence from a national survey of faculty. Journal of the American Medical Association. 277. 1224-28.

Breschi S. and F. Lissoni. 2001. Knowledge spillovers and local innovation systems: a critical survey. Industrial and Corporate Change. 10:975-1005.

Brooks, H. 1994. The relationship between science and technology policy. Research Policy. 25: 477-486.
Caro, J.M.A., I.F. de Lucia and A.G. Gracia. 2003. University patents: Output and input indicators ... of what?. Research Evaluation 12: 5-16.
Cohen, W.M., R.R. Nelson and J. Walsh. 2002. Links and impacts: The influence of public research on industrial R&D. Management Science. 48: 1-23.
Dasgupta, P. and P.A. David 1994. Towards a new economics of science. Research Policy. 23: 487-521.
Etzkowitz, H. and L. Leydesdorff (eds.). 1997. Universities and the Global Knowledge Economy: A Triple Helix of University-Industry-Government Relations. London: Cassell Academic.
Feldman M.P. 1999. The new economics of innovation, spillovers and agglomeration: a review of empirical studies. Economics of Innovation and New Technology, 8: 5-25.
Florida, R. 1999. The role of the university: Leveraging talent, not technology. Issues on Science and Technology. 15: 67-73.
Fontana, R. Geuna, A. and M. Matt 2004. Firm size and openness: The driving forces of university-industry collaboration. In Y. Caloghirou, N. Constantelou, N. Vonortas (eds) Knowledge flows in European Industry: Mechanisms and Policy Implications, Routledge, Forthcoming.
Gibbons, M., Limoges, C., Nowotny, H., Schwartzman, S. Scott, P., and M. Trow 1994. The New Production of Knowledge. Sage: London
Geuna, A. 1999. The Economics of Knowledge Production: Funding and the Structure of University Research. Edward Elgar: London
Geuna, A. 2001. The changing rationale for European university research funding: Are there negative unintended consequences', Journal of Economic Issues. 35: 607-632.
Geuna, A., Steinmueller W.E. and A.J. Salter (eds). 2003 Science and Innovation. Rethinking the Rationales for Funding and Governance. Cheltenham: Edward Elgar.
Gulbrandsen, M. and J.-C. Smeby (2002) 'The external orientation of university researchers and implications for academic performance and management' submitted to Science and Public Policy.
Jaffe, A. 1989. Real effects of academic research. American Economic Review. 79: 957-970.
Laursen, K. and A. Salter. 2003. Searching low and high: Why do firms use universities as a source of innovation? Paper presented at the 3rd European Meeting on Applied Evolutionary Economics, Augsburg, Germany, 10-12 April.
Leydesdorff, L. and H. Etzkowitz. 1996. Emergence of a triple helix of university-industry-government relations. *Science and Public Policy*. 23: 279-286.
Llerena, P. and Meyer-Krahmer, F. 2003. Interdisciplinary research and the organization of the university: general challenges and a case study in A. Geuna, A.J. Salter and W.E. Steinmueller *Science and Innovation: Rethinking rationales for funding and governance*, Cheltenham: Edward Elgar, pp. 80-81.
Louis, K.S., Jones, L.M., Anderson, M.S., Blumenthal, D. and E.G. Campbell. 2001. Entrepreneurship, secrecy, and productivity: a comparison of clinical and non-clinical faculty. *Journal of Technology Transfer*. 26: 233-45.
Mansfield E. 1991. Academic research and industrial innovation. *Research Policy*. 26: 1-12.
Mansfield E. and J.Y. Lee. 1996. The modern university: contributor to industrial innovation and recipient of industrial R&D support. *Research Policy*. 25(7): 1047-1058.
Meyer-Krahmer, F. and U. Schmoch 1998. Science-based technologies: university-industry interactions in four fields. *Research Policy*. 27: 835-852.
Molas-Gallart, J., Salter A.J., Patel, P. Scott A. and X. Duran. 2002. Measuring and mapping third stream activities. Brighton: SPRU-University of Sussex.
Mohnen, P. and C. Hoareau. 2002. What type of enterprise forges close with universities and government labs? Evidence from CIS 2. *Managerial and Decision Economics*. 24: 133-145
Mowery, D., Nelson R., Sampat B., and A. Ziedonis. 2001. The Growth of Patenting and Licensing by U.S. Universities: An Assessment of the Effects of the Bayh-Dole Act of 1980. *Research Policy* 30:99-119.
Narin F. et al. 1997. The linkages between US technology and public science. *Research Policy*. 26: 317-330.
Nelson, R.R. 2001. Observations on the Post-Bayh-Dole rise of patenting at American universities. *Journal of Technology Transfer*. 26: 13-19.
OECD. 2002. Benchmarking industry-science relationships. Paris: OECD.
Pavitt, K.L.R. 2001. Public policies to support basic research: What can the rest of the world learn from the US theory and practice? (And what they should not learn). *Industrial and Corporate Change*. 10: 761-779.

Poyago-Theotoky, Beath, J. and D.S. Siegel 2002. Universities and fundamental research: reflections on the growth of university-industry partnerships. *Oxford Review of Economic Policy*. 18(1): 10-21.

Ranga, M., (2003) 'The exploration-exploitation dichotomy and the impact of environment dynamics on university-industry partnerships', Dphil Thesis SPRU, University of Sussex, forthcoming.

Salter, A.J., D'Este P., Martin B., Geuna A., Scott A., Pavit K., Patel P. and P. Nightingale. 2000. Talent, not technology: Policy funded research and innovation in the UK. Report commissioned by Committee of Vice-Chancellors and Principals (CVCP) and the Higher Education Funding Council for England (HEFCE). http://www.sussex.ac.uk/spru/news/talent.html (last accessed 30/6/2003).

Salter, A.J. and Martin, B.R. 2001. The economic benefits of publicly funded basic research: a critical review. *Research Policy*. 30: 509-532.

Schartinger, D., A. Schibany and H. Gassler. 2001. Interactive relations between universities and firms: empirical evidence for Austria. *Journal of Technology Transfer*. 26: 255-69.

Siegel, D.S., Waldman, D. and A.N. Link. 2002. Assessing the impact of organizational practices on the productivity of university technology transfer offices: an exploratory study. *Research Policy*. 31: 1-22.

Stephan P.E. 2001. Educational implications of university-industry technology transfer. *Journal of Technology Transfer*. 26: 199-205.

Chapter 7

THE NATURE AND THE EXTENT OF THE MARKET FOR TECHNOLOGY IN BIOPHARMACEUTICALS

Ashish Arora, Alfonso Gambardella, Fabio Pammolli and Massimo Riccaboni
Carnegie Mellon University, St.Anna School of Advanced Studies, University of Florence - CERM and University of Florence

Abstract: The biopharmaceutical industry is a typical example of the development of technological collaborations, as well as of technological competition, between larger established firms and smaller high-tech specialist firms (the so-called New Biotechnology Firms – NBFs). These two types of firms perform innovative activities at different stages, with different degree of risk, and with different probability of failure. By using a comprehensive database of 2078 drug R&D projects promoted all over the world during the 1990s, this study assesses the different performance of R&D processes conducted under different governance structures, most notably projects that are fully internalised by the companies vis-à-vis projects developed in collaboration with other firms. Moreover, this study compares the different specialisation and performance of large drug companies with respect to the NBFs. Results show that the established pharmaceutical companies have comparative advantages with respect to NBFs in drug development, while there is no advantage related to scale in drug discovery. Furthermore, NBFs undertake less risky project, which are more likely to fail at earlier clinical stages.

Key words: Research collaborations; R&D Management; Pharmaceutical Industry; Biotechnology

F. Cesaroni et al. (eds.), R&D, Innovation and Competitiveness in the European Chemical Industry, 175-202.

1. INTRODUCTION

Technological collaborations, R&D agreements, licensing and the exchange of technology among independent parties have become common phenomena in recent years. (See for instance Teece, 1986; Kogut, 1988; Hagedoorn, 1992; Hagedoorn and Schakenraad, 1994.) In many industries, and particularly in many high-tech industries, one observes the rise of smaller high-tech firms whose main competence is technology. Typically, these firms supply their technologies through licensing or more elaborated forms of collaboration within full fledged "markets for technology". (Arora, Fosfuri, and Gambardella, 2000.) As argued by Gans, Hsu and Stern (2000), when technological appropriability is strong, these firms tend to act as specialized technology suppliers to the larger established firms which operate in the downstream markets. When technological appropriability is weaker, they tend to appropriate rents from their mastery of the technology by moving downstream, and by competing with the established firms.

Biopharmaceuticals is a quintessential example of the development of technological collaborations, as well as of technological competition, between larger established firms, and a set of new, smaller firms specialized in technology. The story of the New Biotechnology Firms (NBFs) has been told by several authors (e.g. among others, Orsenigo, 1989; Gambardella, 1995). After the foundation of Genentech in 1976, a wave of NBFs were founded in the 1980s, and many others have been founded during the 1990s. The NBFs are specialized in the early stages of drug R&D. Their comparative advantage is in the application of the new biological technologies and the life sciences to drug discovery, and to other areas like agricultural chemicals, food, environment, etc.. In pharmaceuticals, the NBFs often lack the resources to conduct the far more costly clinical trials to obtain the market approval for their new compounds. They also have a comparative disadvantage in large scale commercialization of the new drugs. Clearly, the large firms, and especially the large drug multinationals, keep significant in-house scientific and technological expertise. However, the NBFs often represent a useful source of new compounds and more generally of external technological and scientific competences.

The story of the NBFs raises several questions. Particularly, we lack a systematic assessment of issues like What are the main characteristics and comparative advantages of the larger and smaller firms in the innovation process? Do the larger firms have comparative advantages in downstream R&D

activities, while the smaller firms have comparative advantages in upstream research because of their more suitable organizational structures and systems of incentives (e.g. Arrow, 1983; March and Levinthal, 1993)? Are the smaller high-tech firms taking up more risky projects (e.g. Arrow, 1983; Holmstrom, 1989)? Are the smaller firms providing a greater deal of "exploration" by initiating several new projects which tend to end up earlier, thereby avoiding more costly failures at later R&D stages? Since the life sciences are said to offer a more scientific approach to drug discovery, potential failures can be detected earlier because of the better understanding of the underlying process. One may then expect that the projects of the NBFs, which are more specialized in the life sciences, fail earlier, with implied cost savings from avoiding later failures. Another question is whether collaborative and in-house R&D projects exhibit different probabilities of success. As we shall discuss in this chapter, there are reasons in favor of either one of the two R&D governance structures, and the only empirical study we know on this issue found that internally developed compounds are more likely to succeed (Pisano, 1997).

This chapter employs a comprehensive database of drug R&D projects all over the world during the 1990s. Our sample includes projects that entered at least the first stage of the clinical trials (clinicals I). The database reports systematic information about such projects, like the therapeutic class of the compound which is being tried; the stage (clinicals I, II, III) in which the project failed, or whether it was eventually approved for commercialization; the company that originated the compound, and the one that developed it (which will be the same company in the case of in-house projects); the R&D stage (clinicals I, II, III) in which the compound was licensed; the country in which the trials were conducted, and in which the compound was approved. As we shall discuss in the next section we used only the projects which were terminated (either that failed or that ended up successfully). Our sample is then composed of 2078 projects. The main goal of the chapter is then to exploit the potential of our database to provide a systematic and comprehensive assessment of the technological collaborations in this industry, the role and the comparative advantages of the smaller high-tech NBFs vis-à-vis the established pharmaceutical companies, along with other aspects of the nature and the organization of the innovation process in this sector.

Of course, biopharmaceuticals is peculiar in that, compared to other high-tech industries, its R&D activities unfold "linearly" from discovery to the various clinical tests. This helps our empirical analysis as it avoid the complications associated with the far more convoluted or "chain-linked", as Kline and Rosenberg (1986) put it R&D processes of industries like electronics. Here, for example, we can distinguish fairly neatly between an early, upstream research stage and the downstream ones. However, the very fact that in this industry we can naturally separate features like the upstream vs downstream R&D stages, or the activities that different types of firms (e.g. large and small) do, suggests that it may provide a meaningful perspective to evaluate more generally the nature of these processes. For example, questions like whether or not biotech firms perform more risky projects, or whether they are specialized in upstream research, or about the nature of the in-house vs licensing process and the potential differences in their probability of success, can be informative about similar processes in other industries.

The chapter is organized as follows. The next section presents our database, and it discusses our sample of projects. Sections 3-8 empirically assess various issues about technological collaborations in the pharmaceutical industry. Section 3 looks at the share of compounds developed in collaboration vs those developed internally by the firms, and the role of the NBFs and the larger firms. Section 4 examines whether the NBFs originate more risky technological projects. Section 5 tests the differences in the probability of success of the licensed vs in-house compounds. Section 6 and 7 assess the comparative advantages in upstream research and in drug development of the NBFs and the larger firms. Section 8 examines whether the NBFs projects that failed, tend to fail at earlier research stages. As noted, this may be a consequence of the more scientific content of the technology in which the NBFs are specialized. The concluding section summarizes the overall picture about the nature of the drug R&D process that emerges from our analysis, along with the differential features of the firms involved. The Appendix develops further details about our results, and particularly about the R&D specializations of the larger and the smaller innovators in our sample.

2. THE SAMPLE OF DRUG R&D PROJECTS

Our sample is drawn from the Pharmaceutical Industry Database (PHID) built at the University of Siena. PHID combines several sector-specific proprietary datasets about R&D activity, collaboration and final drug markets with data from public sources as well as companies confidential information and press releases. PHID covers 11418 R&D projects developed by 2262 organizations including: 427 pharmaceutical companies ,1222 biotech firms and 613 universities and other public and private research centers between 1989 and 1999. Twenty-two percent of all the projects in our database were developed in collaboration by two or more partners.

For each R&D project, the database provides the following information:

1) Originator. Firm/institution that launched the R&D project on a new chemical compound with potential pharmacological activity (or known chemical compounds for different pharmacological targets). Typically, it is the holder of the patent on the new compound (or the new indication).

2) Developer. Firm/institution that developed the project. This is the same as the Originator for the in-house projects. For the licensed compounds, the Developer is the licensee who is entitled to develop it further. Frequently the relationship between Originators and Developers is not just a pure licensing contract. In 67% of the cases it amounts to an R&D collaboration with fairly complex contractual and organizational settings. Moreover, 12% of the collaborations are between one Originator and many Developers. In these cases we considered different projects for different Developers.

3) Therapeutic Categories: Projects have been classified according to their targets in terms of likely therapeutic markets. We adopted the ATC (Anatomic Therapeutic Classification) at the 3rd digit level. For example, HIV-antiviral correspond to the ATC3 class J5C.

4) Pharmacological Actions: The main pharmacological activity of the compounds is described in standard terms. Back to the HIV-antivirals example, we can discern proteinase inhibitors from reverse transcriptase inhibitors and other products with different biological targets.

5) Development History: The PHID database monitors the whole development history of the projects starting form the patenting date of the compound (priority and issue), through preclinical and clinical development stages (I, II, III), to registration and final launch on the market. For unsuccessful projects it registered the stage in which they have been discontinued.

6) Collaborations and licensing activity: For the licensed compounds, PHID records the development stage (clinicals I, II, or III) at which the collaboration agreement was signed, and the type of collaboration that was specified by the parties.

7) Country of clinical trials: PHID also reports the countries in which the clinical trials were conducted, and for the successful projects those in which the products were approved. As we shall see below, we only look at the projects in which the Developer is a US or European company. In the vast majority of cases, the companies from these regions run the clinical trials and obtained

approvals in their home country as well. We then took the home country of the company as the country of the trials.

Moreover, for the companies included in our database we collected information about: (a) year of founding; (b) nationality; (c) number of employees; (d) total revenues (health and consolidated); (e) R&D expenditures (health and consolidated); (f) sales in seven countries (USA, Canada, UK, France, Germany, Spain and Italy).

For the purposes of this chapter we extracted from PHID a specific sample. First, we disregarded all the projects that had not yet reached the clinical trials. Firms may differ in their definition of an in-house projects, and they may have different attitudes about releasing news about their internal R&D activities. By contrasts, when they enter the clinical trials, the information about projects is registered and becomes public domain. This also entails a fairly standard definition of what constitutes a drug R&D project.

Second, our sample is composed of all the projects that were terminated. These are all the projects in the database for which we know that the compound was either successfully approved for marketing, or it was discontinued at clinicals I, II, or III. The database does not tell us whether the project actually failed. However, managers in the pharmaceutical industry told us that it is rare that a project that passed some clinical test was not moved to the next stage. This rules out that projects may overcome some clinical tests but were discontinued because of the lack of funds to conduct the later clinical trials. Relatedly, it rules out cases of strategic behavior on the part of the pharmaceutical companies, which may choose not to enter a later trial because they have reasons to believe that the compound may fail at the later stages viz. the expected value of the next trial is not worth its costs, in spite of the fact that the compound passed the earlier stage. In short, the compounds that were discontinued can be treated as genuine failures at that stage of the clinical tests.

Third, we restricted our sample to the projects whose trials were conducted in the US or in one of the 15 countries of the European Union plus Switzerland and Norway. When we selected only the leading European countries, notably the UK, France and Germany, the differences in our empirical results were marginal. Moreover, as we shall see, our results are fairly similar when we restrict our analysis to trials in the US by the US firms. We also confined our sample to projects in which the firm developing the compound is from the US, one of the Europe-15 countries, Switzerland or Norway. By this, we mean companies with their headquarter in one of these countries.

The reason why we did not include trials conducted in other countries is that the European and US trials are likely to be more stringent. This prompted us to eliminate the companies from the other countries. The higher costs of international trials for these companies may prompt them to try in the US or Europe only their best compounds, with implied upward bias in their probability of success. We noted that Japanese companies received many international licenses which were probably licenses for compounds that were already commercialised elsewhere. These were unlikely to be genuine new compounds. To avoid this and related biases we chose to eliminate Japanese companies (and the Japanese trials) altogether. We also excluded all the projects licensed by the established pharmaceutical companies to other companies. Since it is unlikely that an established pharmaceutical firm licenses a product to another firm without commercializing it itself, these are probably not new R&D projects. Most likely, these are projects developed earlier by the licensor, and then tried again, possibly in his home country, by the licensee. Therefore, all the licensed compounds in our sample come either from universities or from NBFs. In addition, we carefully inspected all the remaining collaborations in our sample to verify their genuine R&D content. Our final sample is composed of 2078 projects.

3. THE NBFS AS AN AMERICAN PHENOMENON

It is typically argued from several accounts of this industry that the NBFs are largely an American phenomenon. Our analysis begins by checking the prevalence of the NBFs in the US market, and the extent of the difference between the US and Europe in this matter.

Table 1 reports the percentage of projects in our sample originated by established pharmaceutical companies, NBFs, and universities. The Table also reports the percentage of licensed projects over the total projects in the sample, and the percentage of successful projects, which gives an idea of the orders of magnitude of the unconditional probabilities of success in our sample.

Table 1: Percentage of Projects Originated by Established Pharma, NBFs, Universities; Percentage of Licensed Projects; Percentage of Successful Projects

	Trials in the US/ Developer is a US company	Trials in Europe/ Developer is a European company
Originator:		
Large Pharmas	37.6%	87.4%
NBFs	56.8%	11.5%
Universities	5.6%	1.2%
% of Licensed Projects	7.5%	2.1%
By NBFs	4.8%	1.7%
By Universities	2.7%	0.4%
% of Successful Projects	9.5%	11.0%
N. of observations	965	776

The Table distinguishes between projects developed by US-based companies in the US (i.e. trials conducted in the US), and projects developed by European-based companies in Europe (trials in Europe). The reason why we restricted to projects developed by the companies in their own regions is the one mentioned earlier that companies may develop in other countries only the compounds with higher expected value. However, we also tried several other cases (viz. trials by any company in the US or in Europe, as well as the full sample). The picture that emerges is basically the one offered by the present Table. For example, we found that the European-based companies normally develop their compounds in Europe, and similarly for the US companies. Since our data focus on the projects that were first developed somewhere, this suggests that, as expected, the companies from either regions first develop their compounds in their own region. This also implied that we only added few observations when we looked at the projects by European or US companies in the US, as compared to the US trials by the US firms only, and similarly for the European case.

The Table shows that there are many more projects originatd by the NBFs in the US than in Europe, and there are many more US than European NBFs. If we look at the first column of Table 1 (US trials by US firms), more than half of the projects were originated by the NBFs, compared to 11.5% of the projects in Europe. As noted earlier, the difference in the percentages does not change if one conditions only on whether the companies are US or European (wherever they do the trials), or on whether the trials are conducted in the US or Europe.

Probably as a consequence of the larger presence of NBFs in the US market, the share of licensed projects is higher in the US (or in the case of the US companies). Overall however, the percentage of licensed compounds is small 7.5% in the US and 2.1% in Europe. One could argue that companies may record as internally developed compounds any sort of "small" projects that they have launched or that they may even have just in their mind, or in the mind of their scientists. By contrast, a compound that was developed through a formal collaboration is likely to be a more sizable and realistic project. As discussed earlier, our database is composed however, of projects that entered at least clinicals I. This means that they are all projects of some scale and magnitude, which cannot be mere ideas. As a result, the percentages shown in the Table may be genuine shares of the two types of projects.

From inspecting our data, we also noted that the share of projects in collaborations has increased significantly for projects launched after the mid-1990s. Many of these projects are not in our sample because they have not been terminated. The percentage of projects in collaborations over the total number of projects still under development is higher (19.95%). This suggests that the importance of technological collaborations and the market for technology might have increased in the 1990s, as also argued elsewhere for this and other industries (Arora, Fosfuri and Gambardella, 2000). Finally, Table 1 shows that the probability of success for a compound that enters clinicals I is of the order of 10%.

4. "RISKY" PROJECTS

It has been argued that smaller high-tech firms have a greater propensity to undertake more risky technological projects. For one reason, their informal and less hierarchical organizational structures favor the development of new ideas. Similarly, their internal incentive system is more likely to induce risky ventures than the hierarchical structure of the larger firms.

We asked a pharmacologist to rank the 3-digit ATC class assigned to the projects in terms of their degree of risk. At the 3-digit level, therapeutic classes reflects quite closely to the basic degree of risk and innovativeness of the compounds. For each new product the ATC Scientific Committee decides if it can be classified in a class that already exists. If not, new groups/classes are created for new markets for which these products are indicated. It is possible to identify according to standard pharmacological criteria the classes with a higher innovativeness content. Specifically, our pharmacologist ranked the 3-digit therapeutic class of the projects in our sample on a 0-5 scale. She assigned 0-1 to therapeutic classes that concern pathologies that are fairly well known and for which there are quite a few existing remedies on the market; 2-3 were assigned to therapeutic classes in which the pathology was less well known, and for which there are few remedies on the market; 4-5 cover therapeutic classes in which the pathologies are largely unknown, and there are very few or no remedies (e.g. Alzheimer).

Moreover, since there are projects that span more than one therapeutic class, we took the average score of the classes assigned to the project, and associated that average value to it. In order to keep a fair number of projects per class of risk, we grouped our risk classes in three main categories. We defined R1 to be our low risk class. This includes all the projects with original score between 0 and 1.68; R2 is the intermediate risk class, for projects whose original score was between 1.68 and 3.34; R3 is the high risk class, for projects whose original score was between 3.34 and 5. As a check for the validity of our classification, we computed the mean probability of success of our projects in the three classes. When considering only the projects developed by the US firms from clinical trials in the US, the expected probability of success for R1, R2, and R3 projects are respectively 15%, 9%, and 4%; for projects developed by the European firms in Europe, they are 15%, 12%, and 7%. This suggests that our risk classes do capture the lower probability of success of taking R&D projects to completion as one moves from R1 to R3.

To assess the propensity of different types of agents to launch risky project, we report the cross-tabulation of the number of projects in classes R1, R2 and R3 originated by the NBFs, the established pharmaceutical companies and the universities. Table 2 employs the sample of projects in which the developer is a US firm in the US. Since most of the NBFs are from the US, this ruled out potential biases due to international projects. For comparison, Table 3 reports the same data for the full sample.

Table 2: Number of Projects by NBFs, Large Pharma, Universities by classes of Risk, R1, R2, R3 – Trials in the US/ US developers

Classes of Risk Originators	R1	R2	R3	Total
NBFs	147	306	95	548
Large Pharmas	59	203	101	363
Universities	8	34	12	54
Total	214	543	208	965

Table 3: Number of Projects by NBFs, Large Pharma, Universities by classes of Risk, R1, R2, R3 – Full Sample

Classes of Risk Originators	R1	R2	R3	Total
NBFs	184	394	131	709
Large Pharmas	223	703	348	1274
Universities	15	57	23	95
Total	422	1154	502	2078

The interesting result of Tables 2 and 3 is that the NBFs launch relatively more projects in R1 than in R2 and R3, while the larger firms (and the universities) originate comparatively more projects in R2 and R3. This is true both in Table 2 (US firms doing trials in the US) and 3 (full sample). Using the data in Table 2, the shares of R1, R2, and R3 projects originated by NBFs are respectively 68.7%, 56.4%, and 45.7%. By contrast, the shares of R1, R2, and R3 projects originated by a large company or a university are: 27.6%, 37.4%, 48.6%, and 3.7%, 6.3%, 5.8%.

This result contradicts the view that smaller high-tech firms undertake relatively more risky R&D, and do more exploration in less known technological territories. It also confirms the point made quite a few years ago by Schwartzman (1976) that the large companies are the more typical vehicle for conducting risky R&D in the drug industry. Put differently, these results confirm that the capability of sustaining high levels of technological risk, such as the one that is necessary to cope with major pathologies with no existing remedies, are more likely to be borne by larger companies with established financial assets and commercialization capabilities. At the same time, these results suggest that what the NBFs are more typically doing is to produce existing remedies using the new biological technologies. They do not tap risky business, but they typically use their expertise in the new technology to tap a market with relatively little risk (e.g. diagnostic kits, rare pathologies, etc.).

5. PROBABILITY OF SUCCESS: LICENSED VS. IN-HOUSE PROJECTS

Are the licensed compounds more likely to be successful than the internally developed compounds? There are reasons for arguing both ways. For example, Pisano (1997) maintained that in the case of licensed compounds there could be a classical "lemon" problem. Asymmetric information implies that the licensor is more informed about the potential of her license. The licensees know this, and if there is no way in which the "good" licensors can distinguish themselves from the "bad" ones, the licensees are only willing to pay license fees that are equal to the expected value between the value of a potentially good compound and that of a potentially bad one. But this means that the licensors of good compounds do not realize the value of their license, while the licensors of bad compounds enjoy an extra rent. As a result, it is likely than on the market one observes a greater share of lemons than in the universe of projects as whole. Pisano tests this hypothesis by checking whether the licensed compounds have a lower probability of success than internally developed ones. By using data on about 300 drug R&D projects he finds that internally developed projects have a higher probability of success than projects developed in collaboration with other parties.

But transaction costs may work the other way as well. For example, due to asymmetric information or other imperfections in the market for technology, licensed compounds have higher costs than internally developed ones. The higher costs are shared by the licensor and licensee. As a result, the expected value of the licensed compound must be higher than an equivalent in-house drug. Since we do not have a good measures for assessing the market size of the drug candidates, we test the hypothesis that licensed compounds have a higher probability of success. Another argument is that the licensed compounds have higher probability of success because of the higher efficiency of the R&D process when associated with an effective vertical specialization in the drug innovation process. As suggested earlier, the NBFs can be more efficient in the discovery stages. If so, the compounds licensed by the NBFs (or the universities) are more likely to succeed.

When comparing the probability of success of in-house vs licensed compounds, one also needs to compare the right conditional probabilities. Compounds licensed at a given clinical stage have passed the earlier stages. Their probability of success is then the probability of success conditional on the fact that they passed such earlier stages. As a result, they should be compared with in-house projects conditional upon the fact that they have overcome the same earlier stages. Since our data provide the information about the stage at which the licensed compounds were licensed, we could compute the probability of success for compounds licensed at clinicals I, II and III. Since clinicals I is the basic clinical stage at which the projects in our sample were recorded, we compared the probability of success of the compounds licensed in clinicals I with the probability of success of the internally developed compounds in our sample. We then compared the probability of success of the compounds licensed in clinicals II with the in-house compounds that passed the clinicals I stage. Similarly, we compared the compounds licensed in clinicals III with the internally developed compounds that passed clinicals II.

The results are in Table 4. As we did earlier, we show both the results for the US companies running US trials, and those obtained using the full sample. The two samples show no qualitative difference in the results. Also, to avoid any potential bias in the data, the sample used in this and the next sections does not include the compounds developed by the universities (but it still includes compounds originated by universities and licensed to companies). This reduces our full sample from 2078 to 2036 projects.

Table 4: Probability of Success, Licensed vs In-house Projects (Standard errors in parenthesis)

	US trials/ US developers	Full Sample
Prob.(Success | Compound licensed in Clinical I)	22.0% (3.8)	20.0 (3.0)
Prob.(Success | Internally developed compounds)	6.5 (0.9)	8.8 (0.7)
Prob.(Success | Compound licensed in Clinical II)	72.7 (13.4)	86.2 (8.1)
Prob.(Success | Internally developed compounds that passed clinicals I)	26.5 (3.1)	26.4 (1.8)
Prob.(Success | Compound licensed in Clinical III)	73.3 (12.7)	73.3 (7.2)
Prob.(Success | Internally developed compounds that passed clinicals I and II)	41.5 (4.2)	40.9 (2.4)

Table 4 shows that the probability of success of the licensed compounds is systematically higher than that of the internally developed compounds. This contradicts the earlier finding by Pisano (1997) that licensed compounds are more likely to be lemons. By contrast, it suggests that either licensed compounds need to have a higher expected value to overcome transaction costs, or that licensed compounds are originated by companies that are more efficient in the earlier stages. In section 7 below, we show that the NBFs are not necessarily more efficient originators than the established companies (apart from possibly in class of risk R1). This suggests that it is transaction costs that makes licensed compounds more likely to be successful.

The Table also shows that the difference in the probability of success between licensed and internally developed compounds is quite pronounced. While one cannot exclude that our sample is biased in favor of the collaborative projects, this evidence would be consistent with the view that licensed compounds entail rather substantial transaction costs. As a matter of fact, the manger of an important US drug manufacturer told us that:

"In my experience we are very discriminating when it comes to choosing in-licensing partners, because the deal typically involves royalties on sales (therefore sharing the profits, while assuming almost all the development risks). With the inhouse compounds, we assume the same risks (resource requirements to develop compounds), but are assured that the profits are not shared."

First, this suggests that apart from asymmetric information and the like, such collaborative deals may entail long negotiations before they are concluded. More importantly, as our manager points out, the resource requirements that are necessary to develop a licensed drugs may not be substantially different from those of an in-house project. As the licensees have to share profits while still incurring the bulk of the R&D costs, the expected value of the deal has to be sufficiently high compared to an in-house projects. Given the potential market size of the compound, the higher expected returns are enhanced only if the probability of success of the in-licensed drugs is rather high. Note that the latter argument does not stem from transaction costs, but from the very fact that the licensee has to share the rents while incurring the bulk of the R&D costs in any case. Thus, the need for being discriminating in the licensing process, which may imply long search processes and negotiations (transaction costs), along with revenue sharing, can make the sample of licensed projects a highly selected one, with related implications for the probability of success.

6. ASSESSING DEVELOPMENT CAPABILITIES

Different types of companies may have different comparative advantages in different stages of the innovation process. Most typically, the larger companies may have greater comparative advantages in downstream innovation development compared to upstream innovation activities. This is because downstream development activities are more likely to rely on scale and experimentation, and this favors companies that can mobilize enough resources for these purposes. By contrast, the flexible organisation of the smaller firms may be more suitable to enhance creativity, which is a key asset for the early stages of the research process. In this and the next section we test for the comparative advantages of the larger pharmaceutical firms and the NBFs in different stages of the drug innovation process.

In this section we focus on the development capabilities. In order to do so however, we cannot simply look at our sample of projects. This is because, for in-house projects, the probability of success of the compounds may be affected by the upstream capabilities of our companies, and this implies that we cannot disentangle their specific downstream capabilities. We focus on the licensed compounds. We argue that when the compounds are licensed, the "originator" capabilities are with the licensor. If development capabilities did not matter, we ought to observe that different companies would not have different probabilities of success once the compound is licensed. By contrast, if for licensed compounds there are differences in the probability of success, they can be attributed to differences in development capabilities.

We tested for such differences between the large pharma companies on the one hand, and the NBFs on the other. If the originator capabilities are assumed to be with the licensors, any remaining difference between NBFs and larger companies as licensees ought to be attributed to differences in development capabilities. Although we could do so in principle, we did not test for differences among developers of different nationalities (e.g. European and US firms). As noted earlier, this is because of the potential bias in the types of compounds that the companies of the two region develop in the other region.

Table 5 reports probit estimations in which the dependent variable is whether the compounds were successfully marketed or not. The sample is composed of all the licensed compounds, and we run two separate regressions for the US companies running trials in the US and for the whole sample. Apart from risk class dummies (and a dummy for US trials in the full sample probit), we use a dummy for whether the company is a large pharmaceutical company rather than an NBF. The Table reports both the estimated probit parameters and the implied changes in the probability of success.

Table 5: Assessing Development Capabilities, Probit Estimation – Dependent Variable: SUCC = 1 for successful projects, 0 otherwise

(Sample = only licensed projects; standard errors in parenthesis)

	US trials/US Developers		US or EU trials/ US or EU Developers	
	Estimated parameter	Change in Probability	Estimated Parameter	Change in Probability
Const.	-0.08 (0.37)	-0.03	0.42 (0.33)	0.15
Dummy for R2	-0.45 (0.40)	-0.15	-0.33 (0.26)	-0.12
Dummy for R3	-1.30 (0.50)	-0.43	-0.89 (0.30)	-0.32
Dummy for Large Pharma	0.77 (0.32)	0.25	0.42 (0.23)	0.15
Dummy for trials in the US	–	–	-0.56 (0.24)	-0.20
N.of obs.	71		164	

Table 5 shows that the large pharmaceutical companies are better developers. The result is statistically significant. Other things being equal, for trials conducted by the US companies in the US, the probability of success when the licensee is an established pharmaceutical firm is about 25% higher than an NBF. The result is similar for the full sample, although the difference in probability is smaller (15%).

7. ASSESSING THE ORIGINATORS CAPABILITIES

To assess the originators capabilities, we run probit regressions with the dummy for successful projects as our dependent variable. Table 6 reports the results of four sets of probit regressions. The first two regressions use only the sample of projects developed by the US companies in the US, while the latter two use the full sample. In each of the two sets of regressions, the second regression eliminates projects originated by the universities to focus on the comparison between large pharmaceutical firms and the NBFs. The probit regressions include dummies for risk classes, and the full sample regressions include a dummy for clinical trials in the US. The Table shows both the estimated probit parameters and the implied changes in probability.

Table 6: Assessing Originators Capabilities, Probit Estimation – Dependent Variable: SUCC = 1 for successful projects, 0 otherwise

(Standard errors in parenthesis)

	US trials/US Developers				US or EU trials/ US or EU Developers			
	Estim. Param.	Change in Prob.	Estim. Param.	Change in Prob.	Estim. Param.	Change in Prob.	Estim. Param.	Change in Prob.
Const.	-1.12 (0.14)	-0.18	-1.11 (0.14)	-0.17	-0.99 (0.09)	-0.19	-0.99 (0.09)	-0.19
Dummy for R2	-0.31 (0.13)	-0.03	-0.32 (0.14)	-0.05	-0.16 (0.09)	-0.03	-0.15 (0.08)	-0.03
Dummy for R3	-0.67 (0.19)	-0.10	-0.66 (0.20)	-0.10	-0.49 (0.12)	-0.09	-0.52 (0.12)	-0.10
Dummy for Originator = University	0.83 (0.29)	0.13	–	–	0.88 (0.18)	0.17	–	–
Dummy for Originator = NBF	0.05 (0.13)	0.01	0.05 (0.13)	0.01	0.12 (0.09)	0.02	0.12 (0.09)	0.02
Dummy for trials in the US	–	–	–	–	-0.15 (0.08)	-0.03	-0.14 (0.08)	-0.03
N.of obs.	936		910		2036		1981	

The key result of Table 6 is that as originators, the NBFs are no better than the established pharmaceutical firms. Compounds with NBF as Originators have a mere 0.01 or 0.02 difference in probability of success compared to compounds originated by the larger firms, and this difference is not statistically significant. By contrast, the compounds originated by the universities have a statistically significant difference in probability of about 0.13. The latter result probably reflects the fact that the universities do not have the type of organization and resources that are needed to test new compounds on a large scale. Thus, if a university starts the clinical trials, it is likely to do so because it has a potentially valuable compound. In short, with universities, the type of projects in our sample might be biased in favour of the more successful projects. Clearly, our finding would not either be inconsistent with the hypothesis that universities have better capabilities as originators. Ultimately, both factors probably contribute to the higher observed performance of the universities as originators.

The finding about the NBFs and the large firms needs some further comment. First, the results of the previous section showed that the large firms have higher absolute advantages downstream, while this section showed that they have no advantage as originators. Thus, the large firms have higher comparative advantages in downstream innovation development. This also suggests that scale is important in drug development, while it is not important in drug discovery.

Second, the results in Table 6 do not account for possible differences between the NBFs and the large firms in different types of innovation projects. Table 7 then reports the sample averages of the probability of success for the large firms and the NBFs in our three risk categories R1, R2, and R3. We report the results for both the sample of the US firms running US-based clinical trials, and for the full sample.

Table 7: Assessing Originators Capabilities, by Risk Category (OLSQ Estimation) (Standard errors in parenthesis)

	US trials/US Developers			US or EU trials/ US or EU Developers		
	R1	R2	R3	R1	R2	R3
Const.	0.10 (0.05)	0.08 (0.02)	0.04 (0.02)	0.13 (0.03)	0.13 (0.02)	0.08 (0.02)
Dummy for Originator = NBF	0.06 (0.05)	-0.01 (0.02)	0.00 (0.03)	0.09 (0.04)	-0.00 (0.02)	0.02 (0.03)
Dummy for trials in the US	–	–	–	-0.03 (0.04)	-0.01 (0.02)	-0.05 (0.02)
N.of obs.	205	509	196	405	1097	479

Table 7 shows that the NBFs have no advantage in categories R2 and R3, but they do have an absolute advantage in R1. Although not statistically significant, the point estimate indicates that in R1 the probability of success of a large US pharma in the US is about 0.10, while for an NBF it raises to 0.16. For R2 and R3, large firms have probabilities of success equal to 0.08 and 0.04 respectively. These probably are roughly the same for the NBFs. The same applies if we look at the full sample, rather than the US firms in the US. In the full sample, the difference between the large firms and the NBFs in R1 is even statistically significant. As shown in section 4, NBFs also run relatively more projects in R1. The NBFs are then consistently specialized in the areas where they also show comparatively better capabilities.

8. FAILING AT EARLIER STAGES

Finally, the advantages of the NBFs may not relate solely to the higher or lower probability of success of the drugs that they develop or originate. In drug R&D, clinical trials can be very costly. Hence, other things being equal, one would prefer to discontinue a potentially bad project earlier rather than later. We then check whether given that a project was discontinued, the NBFs tend to discontinue them at earlier rather than later stages.

To do so, we restricted our sample to all the projects that were discontinued. We constructed a variable DISC which takes the values 0, 1, 2 according to whether the project was discontinued in clinical I, II, or III. As usual, we show the results for both US firms in US trials, and for the full sample which includes European firms and European trials as well. Also, to focus on the NBFs vs large pharma comparison, we eliminated from our sample the projects originated by universities. We run multinomial logit regressions using the following variables as regressors: risk dummies, dummy for NBFs as an originator; and for our full sample regressions we added a dummy for trials in the US. The multinomial logit results are in Table 8. Table 9 reports the implied changes in probabilities.

Table 8: Multinomial logit, Dependent Variable: DISC = 0, 1, 2 for projects discontinued in clinicals I, II, III. Sample = Discontinued Projects (Standard errors in parenthesis)

	US trials/US Developers	US or EU trials/ US or EU Developers
1st set of Parameters: Differences b/w DISC=1 and baseline case DISC=0		
Const.	-1.10	-1.04
	(0.29)	(0.18)
Dummy for R2	-0.22	-0.21
	(0.30)	(0.19)
Dummy for R3	-1.01	-0.29
	(0.41)	(0.22)
Dummy for Originator = NBF	-1.48	-1.15
	(0.26)	(0.20)
Dummy for trials in the US	–	-0.37
		(0.16)
2nd set of Parameters: Differences b/w DISC=2 and baseline case DISC=0		
Const.	-0.83	-0.82
	(0.27)	(0.17)
Dummy for R2	-0.66	-0.58
	(0.30)	(0.18)
Dummy for R3	-0.60	-0.45
	(0.34)	(0.20)
Dummy for Originator = NBF	-1.64	-1.52
	(0.26)	(0.21)
Dummy for trials in the US	–	-0.05
		(0.15)
N.of obs.	832	1759
of which		
DISC=0	672	1292
DISC=1	78	225
DISC=2	82	242

Table 9: Estimated Changes in Probabilities for NBF as Originators from the Multinomial logit Model in Table 8 (Change in Probability of failing at Stage I, II, III given that Project was originated by an NBF)

Stage of Project Discontinuation	US trials/US Developers	US or EU trials/ US or EU Developers
Clinicals I	0.22	0.24
Clinicals II	-0.10	-0.09
Clinicals III	-0.12	-0.15

The results show that the projects originated by the NBFs are more likely to be discontinued at earlier clinical stages. As shown by Table 9, the probability that a project originated by an NBF is discontinued after clinicals I is about 0.22 higher than that of a large pharma; the probability that it is discontinued in clinicals I or II is about 0.12 higher for an NBF (0.22-0.10). One has to be careful in interpreting these results. As discussed earlier, a project that was discontinued may not be necessarily one that failed. We know that it was discontinued by the company, and this might be because the company could not raise the financial resources that were necessary to continue the trials. Clearly, this problem is far more likely to occur for the NBFs than for the large pharmaceutical companies. This suggests that a potential source of bias is that the NBFs may discontinue projects earlier simply because they do not possess the necessary assets to continue the project rather than because the project actually failed.

However, projects are normally discontinued because the company does not think that it has good prospects to pass the later stages. An NBF is then more unlikely to discontinue a potentially good project. Put differently, discontinued projects are in any case projects that have lower expected probabilities of success. Thus, one fairly conclude that given our results in Tables 8 and 9, another potential advantage of the NBFs is that they can provide a good deal of exploration in drug research, as they produce information about potential compounds without engaging in the costly donwstream trials.

Another hypothesis is that the greater ability of the NBFs to discontinue their projects earlier could stem from the nature of their underlying knowledge basis and technology. The life sciences are strongly rooted in scientific understanding of the biological phenomena. This provides a predictive capabilities of the performance of new compounds without having to engage in long and costly tests. Similarly, the new computerized drug screening technologies, and related techniques and devices typically used by the NBFs (e.g. combinatorial chemistry), provide a significant ability to "predict the failures". In short, the knowledge basis and technologies of the NBFs encourage the possibility of trying new molecules systematically and discover relatively early whether they are potentially valuable or not, with implied potential for greater opportunities of exploration.

9. CONCLUSIONS

This chapter explored an unusually comprehensive data set of drug R&D projects all over the world. Our data base enabled us to characterize several features of the potential for success of the innovation process in pharmaceuticals. Since this industry is typically characterized by the role of both larger established companies, and smaller high-tech specialist firms the so-called NBFs our analysis enabled us to assess the different performance of R&D processes conducted under different governance structures most notably projects that are fully internalized by the companies vis-à-vis projects developed in collaboration with other companies. Moreover, our data enabled to assess the different specialization and the comparative performance of different types of firms, and particularly of the large drug companies with respect to the NBFs.

Our results can be summarized as follows:

a)The NBFs are largely an American phenomenon. More than half of the drug R&D projects originated in the US are by NBFs, while almost 90% of the drug R&D projects originated in Europe are from established pharmaceutical firms;

b) Collaborative R&D projects are consistently more likely to occur in the US than in Europe. However, in-house projects are a significant majority of the drug R&D projects that entered the clinical stages.

c)The established pharmaceutical companies have comparative advantages with respect to NBFs in drug development (clinical trials).

d) In drug discovery there is no advantage related to scale. Unlike clinical developments, where the large firms seem to have superior capabilities when compared to the NBFs, there is no inherent superiority (in terms of ultimate probability of success of the compounds) between the NBFs and the large firms.

e)The NBFs and the large firms do not seem to have differential capabilities in drug research in therapeutic classes that feature higher degree of risk (R2 and R3). The NBFs however, have higher absolute capabilities in therapeutic classes with low levels of risk (R1). Consistently, they have a higher share of projects in such therapeutic classes.

f) The relatively higher share of NBF projects in R1, as opposed to R2 and R3, suggests that the NBFs do not undertake more risky projects (as noted for instance by the literature which suggests that smaller high-tech companies undertake more risky ventures). More risky projects (R3) are more likely to be undertaken by the larger pharmaceutical companies, which suggests that scale, market power, and the ability to moblise large amounts of resources are key factors in enabling the firms to sustain such higher risks.

g) Finally, other things being equal, the projects originated by the NBFs are more likely to fail at earlier clinical stages. This suggests that the NBFs perform a good deal of exploration without incurring the higher costs of failing at later stages.

10. APPENDIX

More on the "specializations" of the NBFs and the large pharmaceutical companies

The ranking of ATC classes in terms of risk that was introduced in section 4 may be further controlled and specified by looking at an analogous classification used in Pharmaprojects (PJB Publications), one of the leading source of information on drugs in R&D. As a whole, Pharmaprojects contains detailed information over 25,000 compounds and pharmaceutical formulations investigated since 1980, including those currently in research and those whose development has been discontinued. In addition, Pharmaprojects profiles each therapeutic category by defining the pathology in medical terms and providing a review of current treatment, epidemiological and market data, and an analysis of therapeutic trends. For each therapeutic area Pharmaprojects provides also two synthetic indices: the Novelty (N) and Market Size (M) Ratings.

On the one hand, the Novelty rating identifies the most advanced drug development strategies: highly innovative combinations of therapy and mechanisms of action received an high score, whereas the drug with a well-known strategy of action are low-ranked. On the other hand, the market size rating is based on the reported world wide sales for the primary therapeutic category.

Table A1 reports the distribution of R&D projects among the NBFs, the large pharmaceutical companies and the Universities with respect to the Risk, Market Size and Novelty classifications. The risk classification combines the novelty and the market size criteria used by Pharmaprojects. NBFs show a higher degree of innovativeness than large pharmas as they tend to apply new mechanisms of pharmacological action especially to therapies for rare pathologies (e.g. Anemia, Gaucher Disease) and niche markets (Transplant rejection of specific organs). On the contrary, large pharmas tend to be active against "pathologies that affect the masses" in larger markets, characterized by complex clinical trials and more alternative therapeutic drugs already in the marketplace. Table A2 reports the most important products awaiting approval to be launched on the US market for deseases that have the highest need of treatment. With rare exceptions (Biomira's Theratope, Vaxgen's AIDSVAX) these products have been developed by large established pharmaceutical companies in-house or, less frequently, in collaboration with NBFs (Xolair, Visudyne, SnET2).

Table A1: *Share of projects by NBFs and large pharmas according to Risk, Market Size of the potential compounds, Novelty of the chemical structure of the molecule.*

		Low	Medium	High
NBFs	*Risk*	*25,95*	*55,57*	*18,48*
	Market Size	43,23	43,75	13,02
	Novelty	37,10	52,93	9,97
Large Pharma	*Risk*	*17,50*	*55,18*	*27,32*
	Market Size	37,47	45,27	17,26
	Novelty	42,05	50,71	7,24

Table A2: The most important programs in late-stage development (Phase III and Awaiting Approval). FDA Clinical Trials, Oct. 2000

Market	Product Name	Description	Developers	Development Strategy Novelty
Anticancer	Theratope	Therapeutic vaccine for metastatic brest cancer	Biomira Inc.	Established
	Faslodex	Advanced brest cancer	AstraZeneca	2nd, 3rd or 4th Compound
	Arzoxifene	Selective estrogen modulator	Eli Lilly & Co.	2nd, 3rd or 4th Compound
Respiratory	Xolair	First anti-immunoglobulin E (IgE) for asthma and rhinitis	Novartis, Genentech, Tanox Inc.	Leading
	Advair Diskus	Antiasthma	Glaxo Wellcome	New Formulation
	Asmanex	Antiasthma	Schering-Plough Co.	Established
Cardiovascular	Vanlev	Antihypertensive	Bristol-Myers Squibb	Leading
Bacterial infections	Ketek	The first in a new family of antibiotics called ketolides	Aventis	Leading
	Cancidas	Antifungal antipneumocystic agent	Merck & Co.	Established
	Factive	inhibitor of topoisomerases I and IV	SmithKline Beecham	Leading
	Spectracef	orally-active 3rd-generation cephalosporin	TAP Pharmaceutical	Established
Depression	Vestra	inhibits norepinephrine reuptake and blocks Alpha2-receptors	Pharmacia Corp.	Leading
Gastrointestinal	Nexium	the first proton pump inhibitor developed as an isomer	AstraZeneca	Leading
	Zelmac	irritable bowel syndrome	Novartis	2nd, 3rd or 4th Compound
Osteoporosis	ALX1-11	recombinant parathyroid hormones	NPS Pharmaceuticals	Leading
	Forteo	recombinant parathyroid hormones	Eli Lilly & Co.	New Formulation
Ophthalmic	Memantine	glaucoma	Allergan Inc.	Leading
	Xalcom	open-angle glaucoma	Pharmacia Corp.	Established
	Visudyne	age-related macular degeneration	QLT Inc., Novartis	Leading
	SnET2	age-related macular degeneration	Miravant Medical Technologies, Pharmacia	2nd, 3rd or 4th Compound
Diabetes	Glucophage	antidiabetic	Bristol-Myers Squibb	New Formulation
	Starlix	type-2 diabetes	Novartis	Leading
AIDS	Trizivir	combination of Epivir, Ziagen, and Retrovir	Glaxo Wellcome	New Formulation
	Aidsvax	preventive vaccine	VaxGen Inc.	Established

11. REFERENCES

Arora, A., Fosfuri, A. and Gambardella, A. (2000) Markets for Technology: Why Do We See Them, Why Don't We See More of Them, and Why Should We Care, manuscript, Carnegie Mellon University, Pittsburgh.

Arrow K.J., 1983, "Innovation in Large and Small Firms", in Ronen J. (ed.) Entrepreneurship, Lexington Books, Lexington, Ma.

Gambardella, A., 1995, Science and Innovation in the US Pharmaceutical Industry, Cambridge, Cambridge University Press.

Gans, J., Hsu, D. and Stern, S. (2000) "When Does Start-Up Innovation Spur the Gale of Creative Destruction?" NBER Working Paper N.7851, Cambridge MA.

Hagedoorn, J. (1993) "Understanding the Rationale of Strategic Technology Partnering: Interorganizational Modes of Cooperation and Sectoral Differences" Strategic Management Journal, Vol.15, pp.371-385.

Hagedoorn, J. and Schakenraad, J. (1994) "The Effect of Strategic Technology Alliances on Company Performance" Strategic Management Journal, Vol.15, pp.291-309.

Holmostrom, B. (1989) "Agency Costs and Innovation" Journal of Economic Behavior and Organization, Vol.12 (3), pp.305-327.

Kline, S. and Rosenberg, N. (1986) "An Overview of Innovation" in Landau, R. and Rosenberg, N. (eds.) The Positive Sum Strategy, National Academy Press, Washington DC.

Kogut, B. (1988) "Joint-Ventures: Theoretical and Empirical Perspectives" Strategic Management Journal, Vol.9, pp.319-333.

March J. (1991) "Exploration and Exploitation in Organizational Learning", Organization Science, Vol.1 (2), pp. 71-87.

March, J. and Levinthal, D. (1993) "The Myopia of Learning" Strategic Management Journal, Vol.14, pp.95-112.

Orsenigo L. (1989) The Emergence of Biotechnology, London, Pinter Publishers.

Pisano, G. (1997) "R&D Performance, Collaborative Arrangements, and the Market for Know-How: A Test of the "Lemons" Hypothesis in Biotechnology", draft, Harvard Business School.

Schwartzman, D. (1976) Innovation in the Pharmaceutical Industry, John Hopkins University Press, Baltimore MD.

Teece, D. (1986) "Profiting from Technological Innovation: Implications for Integration, Collaboration, Licensing, and Public Policy" Research Policy, Vol.15, pp.285-305

PART 3:

POLICY REGULATION AND THE ENVIRONMENT

Chapter 8

DEVELOPMENT AND DIFFUSION OF ENVIRONMENTAL TECHNOLOGIES

The Role of the European Chemical Industry

Roberta Arduini and Fabrizio Cesaroni
University of Urbino and St. Anna School of Advanced Studies

Abstract: During the last decades, the chemical industry has constantly tried to reduce the environmental impact of its production processes, and to reduce the overall level of pollution, through the adoption of environmentally safe technologies and products. In this respect, the main objective of this study is to assess the role played by the European chemical industry in the development and diffusion of environmental technologies, compared with the US and Japan. Specifically, the study firstly provides a general classification of environmental technologies, which distinguishes between ex-ante and ex-post interventions. Then, on the bases of patent statistics and several case studies, it compares the technological competences of European, US and Japanese chemical companies, and explores the incentives that chemical firms have in developing and adopting clean technologies. The role of specific environmental regulation and of the public opinion (customer demand) is discussed.

Key words: Environmental Regulation; Innovation Incentives; R&D Management; Chemical Industry

1. INTRODUCTION [1]

In recent years, the ecological question has assumed great importance, and the chemical industry has often been accused of being highly responsible for pollution. As a matter of fact, processes of production, distribution and use of chemical products can be very dangerous and some relevant accidents (e.g.

F. Cesaroni et al. (eds.), R&D, Innovation and Competitiveness in the European Chemical Industry, 205-230.

Seveso, Bhopal) have contributed to generating a diffuse suspicion against chemicals. Mainly chemical firms have been highly committed to solving environmental problems.

Since the 1970s, norms and regulations on environmental protection have determined a progressive increase of control measures on the environment, thus creating greater ties for firms. Legislative instruments, particularly "command and control" laws, have become an important constraint for manufacturers.

Besides regulations, another relevant factor influencing the behaviour of firms is represented by the market pressure, and the greater attention that consumers direct towards the environmental impact of products and production processes. In the chemical sector the pressure from public opinion and from the market have both a direct and an indirect influence, through the "supply relationships". Indeed, chemical firms supply many intermediate products to downstream industries, which in turn introduce chemical compounds in their final products. The environmental performance of end products depends on the environmental performance of the entire production line. Therefore, the capability to prevent pollution and the formation of wastes by the chemical industry induces a reduction of pollution in the whole economic system.

In a broad sense, pollution in chemicals is generated from raw materials that are not incorporated in the end products. By reducing undesired by-products, firms not only produce clean or green products, but also obtain more efficient solutions. Hence, the prevention of pollution often determines an increase in process efficiency and productivity. In this respect, among environmental technologies, clean technologies involve both the ecological and the economic dimensions of the firms, and the adoption of such technologies might increase the competitive advantage of the firms, compared to their competitors. These concerns have induced the chemical industry to find new ways to reduce the environmental impact of their production processes.

Against this background, this study aims at analysing the role sustained by the European chemical industry in the development and diffusion of environmental technologies, compared with the US and Japan. Since environmental technologies represent a very heterogeneous category, the analysis has been conducted by making a distinction between end-of-pipe and recycling technologies, on the one hand, and clean technologies and green products, on the other. The first two categories of technologies have been analysed by means of patent data. The latter two categories have been analysed by means of case studies.

Results of this analysis show that the European chemical industry is particularly active in promoting environmental technologies, especially those solutions that prevent the formation of pollution (ex-ante intervention). In order to induce the development of effective technologies, however, a relevant role to governmental intervention through specific regulations is often required by firms. Indeed, competitive concerns per se do not represent sufficient incentives to favour innovation. These factors become relevant in the choice between ex-post or ex-ante solutions, but they come into play only after the decision to develop environmental innovations has been undertaken. They only represent a second-order effect.

The chapter is organised as follows. In Section 2 we recall the main literature on environmental technologies, by identifying the factors that influence firms' innovative decisions. A particular attention is posed on the role of governmental regulation, and on clean technologies, given the impact they play on the firms' competitiveness. Section 3 describes the methodology used in this study. In Section 4, we analyse end-of-pipe and recycling technologies by means of patent data. Contrarily to similar studies, we use two patent datasets (USPTO and EPO patents), as to have the possibility of comparing domestic and foreign behaviours. In Section 5, we analyse clean technologies and green products, by means of case studies. The development of process and product clean technologies is subject to different factors. Hence, this Section makes a distinction between these two types of innovations. Section 6 concludes the paper and explores some policy implications.

2. BACKGROUND

During the last decades, the development of environmental technologies and the adoption of cleaner production processes has become a key element of the innovative activity of chemical companies. The transformations of industrial processes in order to satisfy environmental requirements vary according to the chemical markets. In the case of basic chemicals, the size of the plants, their reduced technical flexibility, the presence of established technological know-how, the need of complementary investments, and the rigidities in terms of raw materials and intermediates to be used, make those transformations more complex. By contrast, these factors play a marginal role in the case of speciality chemicals, where problems usually concern the use of toxic reagents and the purification of final products.

The capability of the chemical industry to reduce the environmental impact of its products and processes has direct consequences on other sectors of the economy. Several chemical products are inputs of downstream manufacturing processes, and cleaner products in chemicals also imply a reduction of pollution in downstream sectors. In some circumstances, it is the need to reduce pollution in these downstream sectors that creates an incentive to the development of environmental technologies in chemicals. In order to satisfy this demand, chemical companies are asked to develop competencies in downstream technologies, in services related to possible uses of chemical products, and in the recycling and disposal of wastes. For example, in order to recover plastic materials, chemical firms have to interact with users and promote an integrated design of materials, which can reduce their environmental impact and facilitate the recovering of subsequent wastes.

Apart from the demand of cleaner products from downstream sectors, the incentives to develop environmental technologies mainly come from the presence of a public pressure and specific environmental regulations, both at the national and international level. As far as public opinion is concerned, consumers' attention to pollution prevention and to the purchase of green products has grown during the last two decades, even though with some differences among countries. The United States and Germany are the two countries where consumers have shown the strongest environmental consciousness. A study focussing on the United States (Ottman, 1993) has shown that US consumers are strongly reluctant to buy products that are believed as dangerous for the environment. For most of the products that have been considered in the survey, firms had to find substitutes within two-three years. In Germany, public pressure for cleaner products has become very strong in the last decades. The number of people describing themselves as "environmentally aware" has grown from 35% in 1982 to 62% in 1989. In 1991, 57% of the population has declared that the preservation of the environment was more important to them than economic growth.

Although the real impact of public opinion on firms' environmental behaviour and strategies is difficult to estimate, it is commonly recognised that people more sensitive to ecological concerns will gear their consumption towards cleaner products, and they will prefer firms which spend their attention to environmental issues. The impact of such consumers is particularly relevant in the chemical industry. Chemical producers have recognised the importance of this public perception. For instance, this was the main rationale for the promotion of voluntary industry initiatives, such as the *Responsible Care* programme. Started in 1985 in Canada, it has suddenly spread to several developed countries. The programme consists of a public commitment to the continuous improvement of health, environmental, and safety performance, and to responding to public concerns about chemical products. Most member companies have implemented some of the codes of *Responsible Care*, and, especially in the US, they have also become increasingly open about their environmental, health, and safety performance. Similarly to the *Responsible Care* programme, other initiatives have been launched in several countries in recent years.

As a direct consequence of a public concern on environmental protection and pollution, governments of almost all the developed countries have introduced some forms of environmental regulations. Policy intervention can be promoted by using different instruments, which differ in their capability to encourage the adoption and development of new, low-polluting environmental technologies. In general terms, these instruments consist of taxes, permits (either allocated for free or by means of auctions), subsidies, and standards (Kemp, 1997). *Taxes* consist of fees for environment exploitation. They provide revenues to governments and represent an incentive for firms to reduce pollution. Each firm looks at its marginal costs and decides the amount of taxes that it is willing to pay. In turn, only those firms whose marginal costs in technology development are lower than the effluent fees will invest in new environmental technologies. *Permits* consist of an emission entitlement limited in terms of quantity and period of validity, allowing the owner to emit pollutants into specified environmental media. Usually, they can be traded among companies. *Subsidies* are grants allocated to companies in order to induce them to invest in environmentally-safe technologies. According to the allocating scheme, they can promote R&D investments, or the acquisition of technologies embodied in capital goods. Finally, *standards* define the amount of pollutants that firms are allowed to emit into the air, water or soil. This instrument is based upon a "command and control" approach, according to which compliance with the standards is checked by the government, and failure to comply is penalised.

Several studies have tried to rank these instruments according to their efficiency.[2] Although results obtained by these studies are often inconsistent, it clearly emerges that policy instruments differ in their ability to encourage the development of different types of technologies, or in the adoption of already available technological solutions. As it will be better exposed in the following section, the reduction of pollution can be promoted by means of alternative technological solutions. At the two extremes, pollution can be reduced at the end of the production process, or it can be reduced at its source, by designing production processes that are intrinsically less pollutant. The two alternatives present different technological complexity. However, they also differ in terms of the influence that they play on firms' productivity and competitiveness. Hence, policy intervention should also take into account these effects together with more general environmental concerns.

2.1 Definition of Environmental Technologies

The concept of environmental technologies defines all technological means and interventions, the aim being to reduce or eliminate pollution and environmental degradation. In such a definition, both purification and treatment plants, and pollution preventing technologies are included. Technologies of the first category do not modify the production technique, they are placed at the end of the production process and aim at transforming wastes into less polluting or not injurious materials. These technologies are usually defined end-of-pipe technologies (or "ex-post", "downstream", "add-on", "au bout de chane" technologies).

A second category of environmental technologies is represented by recycling technologies, which transform wastes in order to re-use them. Recycling can be pursued either within the production process or by means of different solutions.

Finally, a third category of environmental technologies regards clean technologies, which prevent the formation of pollution through an ex-ante intervention. Prevention is widely considered today the preferred method to solve environmental problems. As a matter of fact, both recent US and European environmental policies are oriented towards prevention rather than purification and reclamation. In general terms, technologies aiming at preventing wastes and pollution, at reducing resource inputs and the use of energy, and at using recycled material are considered clean technologies. While end-of-pipe technologies are placed at the end of the production process, clean technologies demand a partial or radical modification of the process. Hence, pollution is avoided at its source or it is reduced, thanks to the recovery of wastes.

With respect to the chemical industry, these technologies aim to reduce, prevent and utilise residues. The term "residue" denotes all components that take part in the chemical reaction, but that are not included in the end product (Wiesner et al., 1995). Because joint production is a constant characteristic of chemical production processes, the existence of residues is very common in this sector.

In addition to clean technologies, green products are those products obtained with recycled or less harmful inputs, or which generate less pollution during their life. Examples of green products are fuel without lead, detergents with low content of phosphorus, and biodegradable plastics.

It is worth noting that the term "clean" or "green" has a relative and not an absolute meaning. Clean technologies and products are not absolutely clean, but cleaner than existing ones. It is also important to note that the boundaries between the various types of technologies are not perfectly defined. For example, new purification plants applying advanced purification processes have the advantage of producing marketable by-products. In this case, the difference between recycling and clean technologies might appear confused.

According to this classification of environmental technologies, clean or green solutions should be preferred to end-of-pipe or recycling technologies. Apart from their technical superiority, the importance of clean technologies stands primarily on the impact they have on firms' productivity (Boninfant *et al.*, 1995; Porter and Van der Linde, 1995; Shrivastava, 1995). Because their adoption implies the modification and re-engineering of the production process, firms can implement such modifications both by making the process less pollutant, and by enhancing the production efficiency. For instance, a lower amount of (polluting) inputs can be used, the production of residues can be reduced, and high-value environmentally-safe products can be obtained. The adoption of clean solutions certainly implies higher costs and higher technical complexity to the firms, compared to different types of environmental technologies. However, contrarily to the traditional view that considers the adoption of environmental technologies a net cost to the firms, they also have an impact on revenues. Indeed, by investing in environmental (clean) innovations, firms can lower the overall production costs, enhance the resource productivity, and strengthen their competitive advantage.

The key point for obtaining such advantages is to distinguish between the hardware component of the technologies, such as pollution control equipments, ecological measurement instrumentation, and cleaner production apparatuses, and the operating methods, such as waste management practices, resource-saving arrangements (Shrivastava, 1995). It is within the management orientation that most of the competitive advantage arising from the adoption of clean technologies can be obtained. For instance, by adopting a total quality environmental management perspective, firms can improve their environmental performance using quality control principles. These principles also allow firms to increase the efficiency of their production process, and obtain products of higher value for their customers. In turn, the adoption of clean technologies results in an increase in firms' productivity, and/or a strengthening of the market advantage with respect to competitors.

Despite the fact that clean technologies affect the firms' competitive advantage, the same firms not always have the right incentives to promote the development and adoption of such technologies. A specific regulation might be needed, for different reasons. First, mainly firms pursue short-term objectives and goals, while environmental technologies usually imply fixed costs that increase the productivity only in the long-run (Simpson and Bradford, 1996). Second, regulation creates pressure that overcome organisational inertia and motivates companies to innovate (Porter and Van der Linde, 1995). It also creates demand for environmental improvements until companies and customers have acquired enough competence to perceive the resource inefficiencies of pollution. Third, in the adoption of a potential productivity-increasing environmental technology a company incurs in costs related to the acquisition of information concerning its efficiency. Once this has been acquired, however, the company cannot fully appropriate the benefits of the information. Other firms will choose whether to adopt that technology according to the value of the information (i.e. the efficiency of the technology) acquired by the former company. In other words, there are social costs in environment technologies development that private companies might choose not to sustain (Farmer *et al.*, 2001).

The critical point in regulation definition, however, is how to favour the development and adoption of clean technologies and not of end-of-pipe or recycling solutions. Regulation, indeed, should create maximum opportunity for innovation, letting firms by themselves understand how to satisfy environmental standards and goals. It should not specify what technology to be used, or set too short deadlines for the adoption of less-polluting technologies, or be ambiguous and hence impose companies high levels of uncertainty. On the contrary, it should permit a more flexible approach, by allowing companies to spend time and resources in the redefinition of their production process in order to satisfy the new environmental needs.

3. METHODOLOGY

End-of-pipe and recycling technologies, and clean technologies have been analysed using different methodologies. As far as end-of-pipe and recycling technologies are concerned, we used patent statistics. Patents have already been used in previous studies in order to meet the same goals. For instance, Lanjouw and Mody (1996) used patent data to study the development and diffusion of environmental innovations in the United States, Japan, Germany and other 14 low- and middle-income countries, over the 1970s and 1980s. For each country, the two authors collected environmental patents from the respective national patent offices. Since environmental technologies are not specifically recognized in the international patent classification (IPC) system, relevant patents have been identified in that study by looking for environmentally-related keywords in the description of the technological classes. Those IPC classes having at least one keyword were included in the analysis. To sum up, nine environmental fields were covered by Lanjouw and Mody's study: industrial and vehicular air pollution, water pollution, hazardous and solid waste disposal, incineration and recycling of waste, oil spill clean-up, and alternative energy.

We considered that methodology as being too restrictive, and eventually decided to use a different approach. First, we classified end-of-pipe technologies in different groups and sub-groups, in order to have a better description of the range of this group of technologies, and in order to better identify the technological specialisation of the countries that we included in our study. On the contrary, we considered recycling technologies as a unique technological class. The final classification is reported in table 1.

Table 1. Classification of end-of-pipe and recycling technologies

A) End-of-pipe technologies		
	A.1) Purification	
		A.1.1) Purification of gases
		A.1.2) Purification of liquids
	A.2) Decontamination	
		A.2.1) Decontamination of soil
		A.2.2) Decontamination of ground water
	A.3) Treatment of wastes	
		A.3.1) Treatment of solid wastes
		A.3.2) Treatment of hazardous wastes
B) Recycling technologies		

Second, we consulted with an expert in the field of chemical technologies, and defined specific keywords for each technological class.[3] Then, we considered as relevant patents only those patents having in the title or the abstract at least one of the keywords identified for each class. Hence, we did not take into account IPC classes. Third, in order to assess the error rate related to the keywords method (i.e. the inclusion of patents not relevant to environmental technologies or not pertaining to the specific technological class), we selected a sample of patents and carefully read the title, abstract and patent description. Hence, we included or excluded specific keywords according to the error rate that they were inducing. The final list of keywords represents a compromise between having a reduced number of perfectly-fitting environmental patents, and having a larger number of patents only loosely connected with environmental technologies. According to the different classes, the error rates range between 5 and 20 percent.

According to this methodology, we extracted information on patent applications filed at the US Patent and Trademark Office (USPTO) and at the European Patent Office (EPO) during the five-years period 1993-1997. Patent data were collected from the database available on the Web, in the case of USPTO patents, and from the Mimosa-Access database, in the case of EPO patents. In both databases, we considered only those environmental patents belonging to agents coming from the US, Japan and 16 European countries.[4]

The possibility to use two different sources of patents allowed us to make cross-comparisons. In each database, the number of domestic patents (US patents in the USPTO database, and European patents in the EPO database) obviously represent the majority. However, the share of foreign patents reveals the interest of foreign assignees (are they firms, institutions of research, government agencies or single inventors) in protecting their innovations in countries different from home countries. This behaviour is usually the result of an evaluation of the market potential for the innovation, and reflects the "quality" of the invention incorporated in the patents. Indeed, it can be reasonably supposed that an assignee will demand the extension of the intellectual property right beyond the national borders only if he or she thinks of possible developments in the foreign market for that invention. In this way the comparison between US patents in Europe, European patents in the US and Japanese patents in the US and in Europe allowed us to characterise technological and competitive forces of these three regions.

Contrarily to end-of-pipe and recycling technologies, patent data were not used in the analysis of clean technologies. Such technologies are more complex than end-of-pipe and recycling technologies, and usually require some complementary organisational change. They cannot be easily incorporated in single patents. Therefore, we preferred to conduct a set of five case studies of chemical companies that were active in developing some successful clean technologies. Specifically, we analysed five Italian chemical manufacturers belonging to large European chemical groups. We chose those companies that agreed to the "Responsible Care" and EMAS voluntary programmes, and which introduced a policy of continuous improvement of the activities related to health, safety and the environment. The first three cases are examples of clean technologies, the last two concern green products.

In each case study, we analysed the forces that induced those companies to undertake the development of more eco-compatible innovations, such as the role of financial incentives, of legislation, and of the public opinion. Then, we tried to give an answer to the following questions: Is the innovative activity conducted through collaboration agreements? Is the collaboration with engineering firms and research organisations relevant to the innovative process?

Do firms patent their innovations? Do they license their technologies? Are clean technologies transferred to other firms and industries? How is the pattern of diffusion and adoption shaped? Finally, we analysed the economic benefits generated by the adoption of clean technologies, and the role of customers in the commercial success of green products.

4. END-OF-PIPE AND RECYCLING TECHNOLOGIES

The innovative rates of the United States and Europe in the end-of-pipe and recycling technologies are very similar. During the period 1993-1997 the average share of environmental patents has been around 1 percent of the total number of patent applications filed at the USPTO and EPO.[5] Only Japan presented a share of environmental patents lower than 1 percent (table 2). However, some differences arise between the two databases. First, the number and the share of patent applications concerning environmental technologies filed at the USPTO are lower than the respective figures related to EPO patents. Second, the share of US environmental patents in the EPO database is lower than the share of European environmental patents in the USPTO database. Hence, the US and Europe (and Japan) present some differences both in their domestic and foreign behaviour concerning end-of-pipe and recycling technologies.

Table 2. Patent applications in end-of-pipe and recycling technologies (1993-97)

	Environmental patents	Total number of patents	Share
USPTO patents			
USA	2705	261514	1.03%
Japan	608	116152	0.52%
Europe	797	77847	1.02%
Other	1558	128089	1.22%
Total	5668	583602	0.97%
EPO patents			
USA	1886	169515	1.11%
Japan	520	65512	0.79%
Europe	2862	203465	1.41%
Other	564	39462	1.43%
Total	5832	477954	1.22%

Note: Other = Patents from countries different from USA, Europe and Japan.

If we analyse in greater detail the different categories of end-of-pipe technologies, the differences between the US, Europe and Japan appear more consistent (table 3). As far as the USPTO patents are concerned, in most classes the largest share of patents is from countries different from the US, Europe and Japan, hence confirming the openness of the American patent system to foreign applications. Apart from these countries, however, the US show the highest shares in decontamination and treatment of wastes, Europe in the purification of liquids, and Japan in the purification of gases.

These results are largely confirmed in the case of EPO patents, apart from the class of treatment of solid wastes where Europe shows a higher share than the US. Taken together, the results from the two databases prove a rather stable technological specialisation of the three countries in different environmental technologies. The case of Japan is particularly interesting since it shows very low shares in most classes apart from purification of gases, where its leadership is particularly strong. In this class, Europe as well shows strong technological competencies, perhaps motivated by the strict regulation on car emissions existing both in Europe and Japan.

Table 3. Patent applications in end-of-pipe technologies (1993-97)

	Purification		Decontamination		Treatment of wastes	
	Liquids	Gases	Soil	Ground Water	Solid	Hazardous
USPTO patents						
USA	0.220%	0.027%	0.086%	0.020%	0.034%	0.039%
Japan	0.076%	0.241%	0.004%	0.000%	0.004%	0.000%
Europe	0.221%	0.132%	0.022%	0.010%	0.015%	0.001%
Other	0.313%	0.048%	0.064%	0.030%	0.054%	0.040%
Total	0.212%	0.088%	0.056%	0.017%	0.030%	0.026%
EPO patents						
USA	0.225%	0.043%	0.070%	0.022%	0.028%	0.031%
Japan	0.150%	0.328%	0.015%	0.000%	0.006%	0.000%
Europe	0.357%	0.217%	0.046%	0.010%	0.029%	0.013%
Other	0.428%	0.155%	0.048%	0.005%	0.041%	0.015%
Total	0.287%	0.165%	0.050%	0.013%	0.026%	0.018%

Note: Share of end-of-pipe patents on the total number of patents by countries.

As far as recycling technologies are concerned, the overall number of US patents in these technologies (both in the USPTO and EPO databases) is greater than those of Europe and Japan (table 4). However, if we look at the propensity to patent in recycling technologies, Europe shows a higher share of recycling patents than the US in both databases. In other words, compared to different technological fields, the effort sustained by Europe to recycling technologies is greater than the effort sustained by the US to the same technologies, even though the difference is not very relevant. On the contrary, Japan shows a major weakness in this field, and the distance from both Europe and the US appear much more relevant.

Within the different European countries, Germany resulted to be the leading innovator both in the field of end-of-pipe and recycling technologies. Only in recycling technologies France and the UK have relevant shares of patents, even though lower than Germany. This result was partially expected, given its economic and industrial structure and the leading position it plays in Europe. However, the "distance" between Germany and the other countries denotes a German thematic specialisation in the field of end-of-pipe and recycling technologies that goes beyond its economic or technological superiority. Our concern is that part of the German leadership in environmental technologies can be explained by the role that regulation played in that country. Indeed, Germany has adopted the most rigid environmental standards of Europe, and this evidence could suggest that a strong regulation has a positive effect on the development and adoption of environmental innovations.

Table 4. Patent applications in recycling technologies (1993-97)

	Number of patents	Share of total patents
USPTO patents		
USA	1594	0.610%
Japan	230	0.198%
Europe	484	0.622%
Other	855	0.668%
Total	3163	0.542%
EPO patents		
USA	1178	0.695%
Japan	193	0.295%
Europe	1494	0.734%
Other	291	0.737%
Total	3156	0.660%

Note: Share of recycling patents on the total number of patents by countries.

4.1 The Role of the Chemical Industry

In order to evaluate the role played by the chemical industry in developing end-of-pipe and recycling technologies, we have analysed the patenting activity in these technological fields of the largest European and American chemical companies. From the 1997 rank of the Global Fortune 500 list, we have extracted the ten largest companies belonging to the general chemical and petrochemical sectors. Then, we have counted the patent applications that they have filed to the USPTO and EPO concerning end-of-pipe and recycling technologies. Results are presented in tables 5 and 6.

In general terms, the chemical industry shows a greater attention to environmental innovations compared to other industries. For instance, if we compare the composition of the patent portfolio of the 20 companies listed in tables 5 and 6 with those of companies from industries different from chemicals, it emerges that these chemical companies have a share of environmental patents which more than doubles the corresponding share of patents of firms from other industries. This result is probably due to the greater attention that the public pressure and governments place on chemical firms, which make them more aware of environmental concerns.

Another interesting result regards the type of environmental innovation promoted by the leading chemical firms. Both in the case of USPTO patents (table 5) and EPO patents (table 6), the figures related to recycling technologies are clearly higher than the corresponding values related to end-of-pipe technologies. In this respect, these companies show a greater effort in the field of prevention rather than ex-post intervention and cure. Furthermore, while table 4 reveals that the US have a higher number of patents in recycling technologies compared to Europe, in the case of the chemical industry, companies from both countries move along the same direction.

Table 5. USPTO patent applications by the largest chemical companies (1993-97)

	End-of-pipe		Recycling		Total	
USA						
Dow Chemical	6	21%	22	79%	28	100%
Du Pont	6	16%	31	84%	37	100%
Exxon	14	27%	37	73%	51	100%
Mobil	14	24%	45	76%	59	100%
Texaco	6	20%	24	80%	30	100%
Chevron	3	33%	6	67%	9	100%
Amoco	3	43%	4	57%	7	100%
Total	52	24%	169	76%	221	100%
Europe						

Akzo Nobel	3	19%	13	81%	16	100%
BASF	10	26%	28	74%	38	100%
Bayer	7	29%	17	71%	24	100%
Hoechst	6	15%	34	85%	40	100%
ICI	2	18%	9	82%	11	100%
Rhone Poulenc	0	0%	3	100%	3	100%
Eni	1	8%	11	92%	12	100%
Ciba-Geigy	4	44%	5	56%	9	100%
Shell	13	50%	13	50%	26	100%
British Petroleum	0	0%	9	100%	9	100%
Elf Aquitaine	1	100%	0	0%	1	100%
Total	47	25%	142	75%	189	100%

It has to be noted, however, that these companies are large multinational corporations, with production facilities and research laboratories spread all over the world. As a matter of fact, while patents by US companies are always owned by the mother company or by their US subsidiaries, patents by European companies are sometimes made by their US associates. For example, concerning the USPTO, all patent applications filed by Shell and Ciba-Geigy, and some patents by Hoechst, BASF and Akzo belong to US subsidiaries. In this respect, data suggest once again the main role of American organisations in patenting recycling technologies. In turn, this result may reflect the greater pressure of legislation and public opinion in the US, rather than in Europe. These pressures were able to encourage the creation of both a market for environmental technologies (especially recycling), and research competencies in this field. As a result, the existence of positive externalities makes it convenient for European firms to locate environmental research in the US.

Table 6. EPO patent applications by the largest chemical companies (1993-97)

	End-of-pipe		Recycling		Total	
USA						
Dow Chemical	6	13%	40	87%	46	100%
Du Pont	4	9%	41	91%	45	100%
Exxon	8	19%	35	81%	43	100%
Mobil	3	7%	42	93%	45	100%
Texaco	6	25%	18	75%	24	100%
Chevron	2	33%	4	67%	6	100%
Amoco	0	0%	5	100%	5	100%
Total	29	14%	185	86%	214	100%
Europe						
Akzo Nobel	8	36%	14	64%	22	100%
BASF	23	29%	55	71%	78	100%

Bayer	9	18%	41	82%	50	100%
Hoechst	14	33%	29	67%	43	100%
ICI	4	16%	21	84%	25	100%
Rhone Poulenc	3	20%	12	80%	15	100%
Eni	0	0%	11	100%	11	100%
Ciba-Geigy	3	8%	33	92%	36	100%
Shell	0	0%	25	100%	25	100%
British Petroleum	0	0%	11	100%	11	100%
Elf Aquitaine	3	50%	3	50%	6	100%
Total	67	21%	255	79%	322	100%

4.2 The Role of Small, Independent Firms in Developing Environmental Technologies

Apart from large chemical companies, a large fraction of end-of-pipe and recycling technologies are patented by different organisations and agents. In order to assess their role, we have extracted a random and stratified sample of environmental patents and controlled for their assignees.[6] In the case of multiple-assignee patents, we considered them separately. In this respect, the total number of patents is different from the total number of assignees. As a matter of fact, we observed that the number of patents with multiple assignee is higher for Japan than for Europe and the US. In the latter, such patents are about 1,5 percent of the total number of patents, while in the former they are about 12 percent. In Japan, therefore, the collaboration between different agents is more likely.

Different assignees have been classified in six classes:

Firms belonging to chemical groups;

Firms belonging to groups of industries different from the chemical industry;

Independent firms, i.e. firms which do not belong to any industrial group;

Governmental agencies;

Universities and other research institutes;

Single inventors.

Within the first two groups we then made a distinction between mother companies and subsidiaries. In order to distinguish firms' typology and sector, the "Who Owns Whom" 1997 directory (WOW, Dun & Bradstreet) was used. Table 7 shows the results related to the USPTO patents (end-of-pipe and recycling technologies, respectively), while table 8 refers to EPO patents.

Table 7. Classification of assignees (USPTO patents, 1993-97)

	End-of-pipe			Recycling		
	Europe	USA	Japan	Europe	USA	Japan
Chemical sector						
Mother company	10%	8%	9%	19%	17%	11%
Subsidiaries	19%	6%	9%	11%	9%	11%
Total	29%	14%	18%	30%	26%	22%
Other sectors						
Mother company	13%	10%	40%	23%	15%	48%
Subsidiaries	19%	6%	21%	23%	14%	15%
Total	32%	16%	61%	46%	29%	63%
Independent firms	36%	50%	19%	19%	38%	15%
Governmental agencies	3%	8%	0%	0%	3%	0%
Research institutes	0%	9%	0%	5%	3%	0%
Single inventors	0%	3%	2%	0%	1%	0%
TOTAL	100%	100%	100%	100%	100%	100%

As far as USPTO patents are concerned, the US show a high prevalence of independent firms, especially in the end-of-pipe sector. In Japan, most innovations are patented by firms non belonging to chemical groups, and the main role is played by the automobile industry in the end-of-pipe sector. In Europe, innovations are mainly patented by not chemical groups in the recycling sector, and by independent firms in the end-of-pipe. It is worth observing that European chemical groups show higher percentages with respect to the US and Japan, and this pattern highlights the relevant role played by the chemical industry in Europe.

In the case of EPO patents, while Japanese assignees behave similarly to the case of USPTO patents, European and US assignees show some differences. In both countries, the chemical groups are mainly active in recycling, and firms belonging to industrial groups in sectors different from the chemical industry patent more in end-of-pipe technologies. As a consequence, independent firms in the US, and industrial groups from sectors different from chemicals in Europe are the main innovators in end-of-pipe technologies. Furthermore, the role played by single inventors appear interesting. Concerning the USPTO patents (table 7), their presence was scarce and limited to some sectors and countries. On the contrary, within the EPO patents, their shares become relevant. For instance, in the case of end-of-pipe technologies in Europe, they patent more than the chemical industry as a whole.

Table 8. Classification of assignees (EPO patents, 1993-97)

	End-of-pipe			Recycling		
	Europe	USA	Japan	Europe	USA	Japan
Chemical sector						
Mother company	9.5%	6%	6%	18%	24%	10%
Subsidiaries	9.5%	10%	11%	19%	15%	14%
Total	19%	16%	17%	37%	39%	24%
Other sectors						
Mother company	14%	6%	22%	8%	13%	47%
Subsidiaries	18%	15%	50%	13%	9%	5%
Total	32%	21%	72%	22%	21%	52%
Independent firms	28%	45%	6%	31%	24%	14%
Governmental agencies	0%	0%	0%	1%	0%	0%
Research institutes	0%	6%	0%	1%	2%	0%
Single inventors	21%	12%	6%	9%	14%	10%
TOTAL	100%	100%	100%	100%	100%	100%

Summing up results from the analysis of USPTO and EPO patents, some considerations emerge. First, US independent firms show a technological specialisation in the end-of-pipe sector, for they have high shares of both domestic and foreign patents. The same can be said by referring to European independent firms, but only when we consider the USPTO patents (i.e., foreign patents). On the contrary, no specialisation emerges by considering domestic patents. It is our concern that most of these firms are active in the sector of "environmental industry", that is in the supply of environmental services and technologies to downstream firms operating in several industries (chemicals and other sectors). This process is particularly apparent in end-of-pipe technologies. Indeed, as a consequence of the introduction of environmental regulation, many firms specialised in the supply of end-of-pipe technologies specifically designed to satisfy the legislative standards.

Second, chemical groups are more oriented towards recycling technologies. They seem more interested in patenting preventive solutions, which are characterised by a greater technological effectiveness and a higher economic efficiency. This result emerges by observing both USPTO and EPO patents. In general terms, however, chemical groups are active in the whole environmental sector.

Third, the Japanese interest on environmental technologies is mainly limited to the end-of-pipe sector. Most innovations in this sector are patented by non chemical groups. Among those, an important role is played by the automobile sector.

Finally, universities, research institutes and governmental agencies have limited shares of environmental patents. Highest values are shown by the US in end-of-pipe technologies.

5. CLEAN TECHNOLOGIES AND GREEN PRODUCTS

More than end-of-pipe or recycling technologies, clean technologies represent the main building block for a sustainable development, as they are more effective than end-of-pipe technologies in preventing pollution and the formation of wastes. Moreover, as we have already discussed, they play a critical role in defining the competitiveness of firms dealing with growing legislative awareness and increasing pressures from the public opinion.

As a matter of fact, during the last decades the chemical industry had to consider human health and environmental impact of new chemical compounds and manufacturing processes. It developed new chemical low-toxicity products, and new processes and syntheses by removing or limiting the use of toxic reagents, producing minimal waste, and requiring minimal energy. All these factors gave rise to the so called *green chemistry*.

In order to analyse the role of chemical firms in developing clean technologies we have analysed five case studies of chemical companies that were active in developing seven successful clean technologies and three green products. The five cases have been chosen among the Italian chemical manufacturers belonging to large European or American chemical groups that agreed to the Responsible Care and EMAS voluntary programmes. The name of the five companies and the environmental projects that we have analysed are listed in table 9. Table 10 reports a synthetic description of the environmental and economic benefits obtained with the adoption of such technologies.[7]

These case studies highlight the great importance that the economic dimension and the possibility to gain economic benefits together with environmental improvements have in the development of environmental innovation. In general terms, each innovation aims at increasing firms' competitiveness, even though some differences emerge among the different cases, and especially between process and product technologies.

Table 9. Case studies of clean technologies and green products

Company	Project
Enichem (ENI)	Dimethylcarbonate (DMC) Rumene Cyclohexanone ammoximation technology
Ciba Specialty Chemicals (Novartis)	Optimisation of gas emissions with solvent recovering and energy production Recycling of waste water
Lonza Intermediates and Additives (Algroup)	ALMA process Optimisation of the phthalyc anhydride production process
I.C.I. Italia (Huntsman ICI "Italian Operation")	Waterlily Vac Pac
Mapei	Ecological adhesives

Table 10. Environmental and economic benefits

Project	Environmental benefits	Economic benefits
DMC	low toxicity and low environmental impact elimination of toxic and polluting chloride materials	creation of not-dangerous and not-polluting products technology and market leadership
Cumene	elimination of toxic catalyst catalyst regenerating increased security of the system management	reduction of reagent specific consumption reduction of maintenance and disposal costs
Cyclohexanone	elimination of NOX and SO2 emissions and of toxic reagents	reduction of investment, management and maintenance costs (process simplification)
Gas emissions	reduction of gas emissions solvent recovery	solvent and energy recovery reduction of purification costs
Waste water	reduction of waste water	reduction of water consumption
ALMA	gaseous emission reduction aqueous effluent reduction process water reduction	use of cheaper raw materials energetic recovery lower investment costs
Phthalyc anhydride	reduction of liquids to be disposed	productivity increase more pure products energetic savings
Waterlily	recyclable product	diversification strategy
Vac Pac	elimination of toxic agents	diversification strategy
Ecological adhesives	low emissions of volatile substances	diversification strategy

In the case of process technologies, their introduction can be motivated either to increase internal competitiveness, or to sell and license that technology on the market for technologies. In the first case, the firm is usually inclined to obtaining a technical improvement of its production processes, and aims to exploiting the benefits of such innovations internally. This decision limits the diffusion of these technologies to other firms or sectors. At the same time, the diffusion of the innovation is also hindered by technical reasons, because of the high technical specificity and contextual knowledge related to the development of the technology. Often these innovations are introduced in order to reduce the pollution generated by the traditional production processes, as a response to existing or incoming regulations. Having to solve an environmental problem, the firm also tries to obtain an economic benefit, by adopting preventive solutions.

On the contrary, when the technology is developed with the aim of selling it in the market, it usually refers to radical innovations. The characteristics of the new production process are usually linked both to environmental and economic benefits. It may occur that the diffusion of such a technology is connected to economic reasons, while the environmental element is connected to the existing (or expected) regulation.

Furthermore, the development of process innovations usually requires the establishment of collaborative agreements. In some cases, firms collaborate with other firms or research laboratories belonging to the same industrial group. In other cases, firms collaborate with external companies or universities. Indeed, the development of environmental process innovations often requires very specialised capabilities and skills, which are hardly possessed by a single chemical firm.

Also in the case of the development of product innovations it is possible to distinguish two possibilities. Indeed, the new product can be directed either to the consumer market, or it can become an intermediate input to be used by other firms in their production process. In the first case, the economic success of the product directly depends on the market demand, or on the existence of a specific regulation that imposes that product in the market. In terms of strategic behaviour, by developing environmental product innovations, the firm aims to exploit specific niches of the market.

In the second case, the impact of consumer demand is less direct, but it still remains important. The main actor in this case is the customer-firm, whose attention paid to the environmental aspects, both regarding its own image and the willingness to exploit some ecological niches of the market, has a direct influence on its choices. The situation in this case presents a stronger complexity. In many cases, the company which introduces a new environmental product also has to "invent" the applications for its use. This occurs particularly in the chemical industry, which often supplies intermediate goods to other industries.

Therefore, the success of a green product depends on several elements. It is firstly linked to the availability of the consumers to buy cleaner products despite their higher prices and, sometimes, their worse technical characteristics. Unfortunately, this availability is rarely sufficiently articulated.

The successful development of clean technologies and green products is also linked to the role of governments, which can intervene through the introduction of specific regulations or thorough the promotion of a demand for green products. If the legislation denies the use of certain products (i.e. CFC), then companies are required to substitute them with other products, even though the new product are more expensive. In this case, the consumers' choice is forced by legislative concerns. Governmental intervention might also consist in the creation of a demand for a certain product. In absence of both these forces, besides environmental characteristics, the introduction of a cleaner product represents a situation of success for the firm only if the products results to be cheaper than existing competing products.

Many of the cases analysed in this study, were the direct or indirect response to the adoption of the EMAS regulation. EMAS is considered by the chemical companies an effective means to stimulate environmentally-related innovations. Indeed, it forces companies to make an analysis of the production processes and to individuate any possibility for improvement. However, it is considered an expensive instrument, especially for small firms. In Italy, the compliance to this rule is mainly caused by the inefficiency of the Italian bureaucracy. As a matter of fact, companies which adhere to EMAS obtain a certificate that gives proof of the adjustment to the normative and an effort to constant improvement. By adopting this approach, firms hope to reduce problems which are mainly connected to bureaucratic reasons. The high costs that firms incur in adopting the EMAS initiative, however, often require further incentives to make it economically effective.

The promotion of initiatives, such as prizes for cleaner technologies, are considered useful to improve the environmental image of the firms, to strengthen a culture oriented to the sustainable growth, and to improve the interaction with the surrounding population. It is worth noting that the firms analysed in our survey belong to large chemical groups in which the environmental policy has been adopted and consolidated for a long time. As a matter of fact, all these companies also adhere to the *Responsible Care* initiative.

6. CONCLUSIONS

In analysing the way in which the European chemical industry contributes to the development and diffusion of environmental technologies, compared to the respective positions of the US and Japan, one of the first results of this study was that the United States patent in the environmental technologies more than Europe, and Germany patents more than all the European countries. The higher innovative rates found for Germany and the US might be linked to the different governmental regulations and public pressures that the two countries face. As a matter of fact, the United States have faced environmental problems through very strict standards, and Germany has adopted the most rigid standards of Europe. Moreover in these countries, public opinion has played an important role in influencing the environmental policy and behaviour of firms. These results could bring about the consideration that rigid environmental standards and strong public pressure have a positive influence on the development and adoption of environmental innovations.

Surely, existing studies have never explicitly shown the existence of a positive correlation between strict environmental regulations and innovative rates, since strict environmental standards can also have a negative effect on other types of R&D investments. Furthermore the environmental regulation can impose significant costs to industry, by lowering its competitiveness. However, the results of our study support the idea that a strict regulation is needed. Obviously, other instruments supporting firms in the adoption and development of preventive solutions have to be implemented as well, along with regulation.

To clarify this matter, it may be useful to summarise the main results of this study. First, the influence of regulation on the innovative activity is suggested by the clear prevalence of the US in the class of "hazardous wastes". The US presents a wide regulation about hazardous wastes, while a similar system of responsibilities does not exist either in Japan nor in Europe.

Second, some European chemical firms prefer to patent environmental innovations in the United States. This pattern shows that a greater market for environmental technologies exists in the US, where relevant technological competencies in this field are concentrated.

Third, the upsurge and development of an environmental industry is closely linked to the introduction and evolution of environmental regulation. Indeed, if the main innovative actors in this sector are firms specialised in the supply of environmental technologies, then they will pursue innovation only if a wide diffusion can be achieved. Clean technologies have developed slowly also because very often the environmental legislation required a too short fulfilment time and the adoption of the already available technologies. But this enforcement represented an indirect pressure to adopt end-of-pipe technologies, and discouraged the development of clean technologies.

Fourth, the analysis of the case studies has shown that the innovative behaviour of firms has to satisfy economic constraints, and that environmental issues are conditioned by the search for technology efficiency and effectiveness. At the same time, when the government regulation defines precise standards, and imposes precise prohibitions, these are clear targets that have to be considered in the innovative activity. They represent a clear reference point, especially when standards are imposed at an international level, and are kept fixed for a sufficiently long time. Hence, innovative problems encountered by firms are not within the regulation itself, but they are related to the time needed to satisfy the standards according to the regulation. If this time is too short, firms are forced to develop end-of-pipe technologies, even though no productivity increase can be reached by adopting these technologies.

Finally, self-regulation may represent a useful instrument. However, its costs can be sustained only in the presence of a strict regulation and/or a high pressure by customers or by public opinion. When these two conditions are not satisfied, there are no incentives for firms to allocate resources on environmentally-safe solutions.

It is worth recalling that the chemical companies (both from the US and Europe) patent more in clean technologies rather than end-of-pipe. Furthermore, in Europe, the chemical industry is the leading innovator in environmental technologies, compared to different actors (other industries or organisations). This result suggests that European policy makers should focus their policies on the chemical industry, also because this industry supplies many inputs to several downstream production processes. In turn, an intervention in the chemical industry would produce beneficial effects on the downstream sectors.

As far as clean technologies are concerned, the small number of licenses related to technologies that our case studies has revealed highlights the problem of their diffusion. We believe that what reduces the transferability of such technologies is not their intrinsic technological complexity. Solutions to reduce or overcome this problem can be found. For instance, the presence of patents may represent an organisational tool to ease up the diffusion processes. On the contrary, the main limits to the transfer of clean technologies are related both to the demand-side, and to the difficulty to recognize the competitive gains that clean technologies eventually induce.

Summing up, in order to favour the development and diffusion of clean technologies rather than traditional end-of-pipe technologies, a policy intervention is required. Competitive concerns, and the possibility to increase internal productivity as a result of the introduction of such technologies, are important effects, but they represent only a second order condition. To be effective, an environmental policy must not only fix low pollution standards, but it also has to promote the transition towards cleaner production processes. Different actions can be undertaken to meet this goal (Lundvall and Borras, 1997), such as the establishment of standards in an interaction between users and producers, or the formation of markets for green products by using procurement policies which involve private and public users, or, in general, the co-ordination between environmental policy, innovation policy and general economic policy.[8]

NOTES

[1] We would like to thank Davide Fratini, Alfonso Gambardella, Walter Garcia-Fontes, Anna Morgante and the participants in seminars at University Pompeu Fabra (Barcelona), University Carlos III (Madrid) and University of Stuttgart for helpful comments and useful suggestions on earlier drafts of this work. We also thank Maria Grazia Ungano for her helpful assistance. The usual disclaimer applies.

[2] Among others, see Downing and White (1986), Milliman and Prince (1989), Jung et al. (1996), Gersbach and Glazer (1999), and Requate and Unold (2003). For a review, see Jaffe et al. (2000).

[3] The list of keywords used in this study is available on request.

[4] The 16 European countries are: Belgium, Denmark, United Kingdom, Finland, France, Germany, Greece, Ireland, Italy, Portugal, Spain, Holland, Austria, Sweden, Switzerland, Norway. It is to be noted that while in the USPTO database the patent nationality consists in the nationality of patent assignee, in the EPO database the nationality corresponds to that of the first country in which the assignee has applied for. Usually, however, such country corresponds to the assignee's nationality.

[5] In table 1 and in most of the tables that follow, the patenting behaviour of a given country in a specific technological class is expressed as a percentage of the total number of patents of the same country (in any technological class) in either the US or the European patent offices. Because the total number of patents varies across countries, normalisation by total patents allows to make comparisons. Therefore, the shares reported in the tables express the propensity to patent in a specific technological class by each country. A similar reasoning has been used in the study by Lanjouw and Mody (1996).

[6] The sample was composed of 10 percent of end-of-pipe and recycling patents identified in the previous analyses. They were proportionally subdivided among countries, and between recycling and end-of-pipe technological classes.

[7] A complete description of the cases analysed in this study can be found in Arduini and Cesaroni (2001).

[8] As it has been emphasised, "Government technology procurement occurs when a government agency places an order for a product or system which does not exist at the time, but which could (probably) be developed within a reasonable period... government technology procurement is an interesting instrument because it may be part of strategy to open up new technological paradigms; it should be more extensively used as a means of enhancing the exploratory side of the innovation process, with specific social goals, like low energy consumption assets and environmental objectives" (Lundvall and Borras, 1997).

7. REFERENCES

Arduini R. and Cesaroni F., 2001, "Environmental Technologies in the European Chemical Industry", *LEM Working Paper* 2001/09, St. Anna School of Advanced Studies, Pisa (Italy).

Boninfant B.C., Arnold M.B. and Long F.J., 1995, "Gaining Competitive Advantage Through Environmental Investments", *Business Horizons*, 38(4): 37-47.

Downing, P.B. and White, L.J., 1986, "Innovation in Pollution Control", *Journal of Environmental Economics and Management*, 13: 18–29.
Farmer A., Kahn J.R., McDonald J.A. and O'Neill R., 2001, "Rethinking the Optimal Level of Environmental Quality: Justifications for Strict Environmental Policy", *Ecological Economics*, 36: 461-473.
Gersbach H. and Glazer A., 1999, "Markets and Regulatory Hold-up Problems" *Journal of Environmental Economics and Management* 37: 151–164.
Jaffe A., Newell R., and Stavins R., 2000, "Technological Change and the Environment", *NBER Working Paper* 7970, NBER, Cambridge (MA).
Jung C., Krutilla K. and Boyd R., 1996, "Incentives for Advanced Pollution Abatement Technology at the Industry Level: An Evaluation of Policy Alternatives", *Journal of Environmental Economics and Management*, 30: 95–111.
Kemp R., 1997, *Environmental Policy and Technical Change*, Edward Elgar, Cheltenham.
Lanjouw J.O. and Mody A., 1996, "Innovation and International Diffusion of Environmentally Responsive Technologies", *Research Policy*, 25: 549-571.
Lundvall B-A. and Borras S., 1997, *The Globalising Learning Economy: Implications for Innovation Policy*, European Comission, Luxembourg.
Milliman S.R. and Prince R., 1989, "Firms Incentives to Promote Technological Change in Pollution Control", *Journal of Environmental Economics and Management*, 17: 247–265.
Ottman J.A., 1993, *Green Marketing: Challenge and Opportunities for the New Marketing Age*, NTC Business Books, Lincolnwood (IL).
Porter M. and Van der Linde C., 1995, "Green and Competitive", *Harvard Business Review*, 73(5), 120-134.
Requate T. and Unold W., 2003, "Environmental policy incentives to adopt advanced abatement technology: Will the true ranking please stand up?", *European Economic Review* 47: 125-146.
Shrivastava P., 1995, "Environmental Technologies and Competitive Advantage", *Strategic Management Journal*, 16: 183-200.
Simpson R.D. and Bradford R.L., 1996, "Taxing Variable Cost: Environmental Regulation as Industrial Policy", *Journal of Environmental Economics and Management*, 30: 282-320.
Wiesner J., Christ C., et al., 1995, "Production-Integrated Environmental Protection", in Gerhartz W. and Elvers B. (eds.), *Ullmann's Encyclopedia of Industrial Chemistry*, Vol. B8, Wiley-VCH, 5th Edition.

Chapter 9

IS BIG BUSINESS GOOD FOR THE ENVIRONMENT?

Empirical evidence from German chemical firms

Frank Becker and Frank C. Englmann
University of Stuttgart, Institute of Economics and Law

Abstract: The objective of this paper is to investigate whether large firm size is conductive to a higher level of inventive activities in the field of water benign process technologies in the German chemical industry during the 1990s. By using patent applications as indicator for inventive activities, our empirical results indicate that with respect to both end-of-pipe and recycling technologies especially very large firms including BASF, Bayer and Aventis (former Hoechst) predominate invention. Except Solvay and Henkel within the portfolio of patent applications of very large firms, patent applications in the field of end-of-pipe technologies play a less important role than patent applications in the field of recycling technologies. Hence, the most important patent applicants focus on a more sustainable technological pathway in protecting the environment during the 1990s.

Key words: Environmental Regulation; Innovation; Chemical Industry

1. INTRODUCTION

The purpose of this paper is to analyse whether inventive activities in the field of water benign process technologies vary with the size of German chemical firms.

We focus on the *chemical industry* because it has been heavily criticised by different societal groups in Germany and elsewhere. Accidents and waste which result from manufacturing, transporting, using and disposing chemical

F. Cesaroni et al. (eds.), R&D, Innovation and Competitiveness in the European Chemical Industry, 231-246.

substances endanger both human health and the environment. Therefore, it is interesting to know the extent of efforts of chemical firms to reduce pollution and to know whether these efforts depend on the size of firms. The rationale for our expectation that firm size is an important determinant for environmentally safer process inventions will be given in the following section.

Furthermore, we concentrate on one specific type of environmental pollution, namely *water pollution* caused by the production of chemical substances. Water pollution is one of the main sources of industrial pollution in the chemical sector. Waste water is produced by cooling chemical processes, by cleaning equipment and pipes, by the employees of chemical companies, and by joint products in chemical reactions. Waste water resulting from joint products is an idiosyncrasy of the chemical industry (Müller-Fürstenberger, 1995): The conversion of a starting material and at least one reaction partner (supported by the addition of further substances such as solvents, some auxiliary substances and a catalyst) into the desired chemical substance is *necessarily* accompanied by the production of some undesired joint products which may contaminate water.

The rest of this chapter is organised as follows. Section 2 gives the rationale for analysing the relationship between inventive activities in the field of water benign process technologies and firm size in the chemical industry. Furthermore, the validity of the patent application indicator will be briefly discussed. Section 3 systematises water benign process technologies. The established technological categories constitute the basis for selecting patent classes and keywords. Section 4 describes the sample of chemical firms which were selected to analyse the relationship between patent activity and firm size. Section 5 presents the empirical results based on descriptive statistics. Section 6 concludes.

2. INVENTIVE ACTIVITIES AND FIRM SIZE

The issues of firm size, market structure, monopoly, competition and technical progress where brought into mainstream economics by Joseph A. Schumpter and have been widely investigated by economists for a long time (see e.g. Scherer, 1965; Pavitt, Robson, Townsend, 1987; Symeonidis, 1996).

The previous investigations suggest that the question of how firm size relates to the ability to invent and innovate can not be answered in a generalised fashion. Among others, industry, technology, and history matter (Freeman and Soete, 1997, p. 240).

However, with respect to chemical processes historical investigations showed that large firms predominated both invention and innovation (Freeman and Soete, 1997, pp. 85-136 and 227-241; Symeonidis, 1996, p. 38) which indicates a trade-off between static and dynamic efficiency.

One important reason for that relationship is the existence of economies of scope in the production of invention and innovation because of positive spillovers between different R&D projects within a firm. But even more importantly, large chemical firms have sufficiently large sales to cover the fixed costs of R&D projects which are very high in industrial chemistry (Freeman and Soete, 1997, p. 234). Thus, we expect a similar pattern of invention and innovation with respect to a particular type of process technologies – technologies that protect water resources.

In order to analyse the relationship between firm size and invention we use patent applications. Patent applications are a rather good indicator for inventive output in chemical industry because patents represent technological ideas with an inventive step and the propensity to patent inventions is high in the chemical sector (Freeman and Soete, 1997, p.113). Furthermore, by using patent applications we avoid any subjective assessment of inventions.

3. SELECTION OF PATENT CLASSES AND KEYWORDS

3.1 Water benign process technologies

Before selecting patent classes and keywords for our empirical analysis we have to systematise water benign process technologies. Usually two classes of water protection process technologies can be distinguished: (i) *end-of-pipe* (or additive) technologies which are added downstream to chemical processes. They prevent the direct release of hazardous emissions into the environment and can be divided into three subclasses: one treating waste gas, one treating waste water and one treating solid waste, such as sewage sludge. Here, the last two subclasses are of interest (see figure 1). (ii) *production-integrated* (or clean) technologies which in turn can be divided into two subclasses: (iia) technologies that reduce emissions at source by changing input materials or processes. (iib) recycling technologies that return valuable material in polluting streams to the original process or to another process of the plant. The recovery of material results in less hazardous waste streams. Here, both subclasses of technologies are of interest.

End-of-pipe and production-integrated technologies constitute two separate technological paradigm in Dosi's sense, i.e. they embody specific outlooks, specific definitions of relevant problems and specific patterns of enquiry (Dosi, 1982, p. 148).

Figure 1. End-of-pipe and production-integrated technology to reduce waste water.

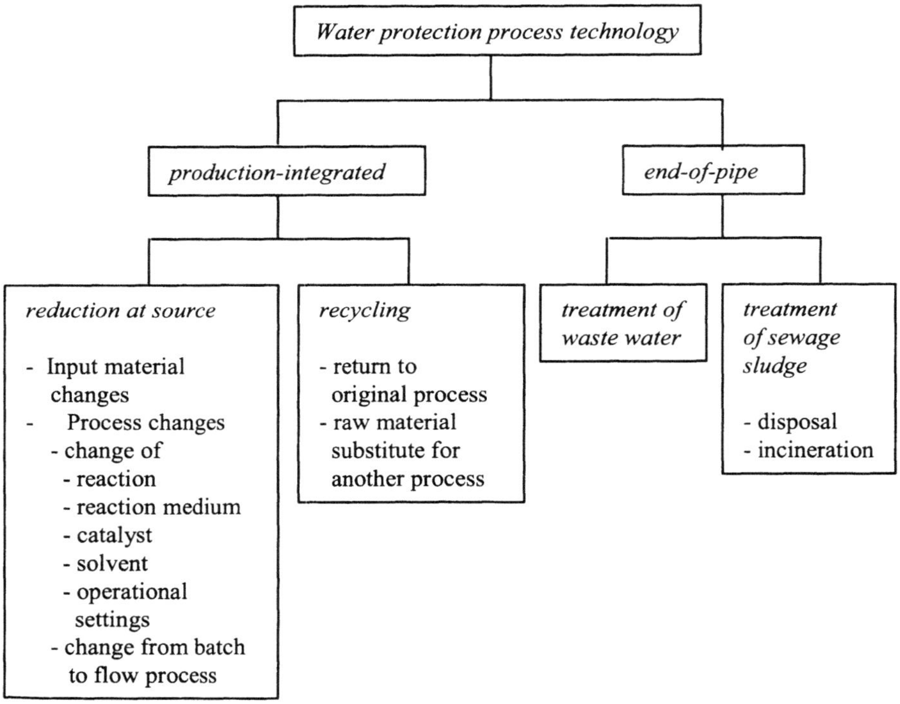

Source: The figure adapts Hohmeyer and Koschel, 1995, p. 6

End-of-pipe technologies modify emissions in order to make them less polluting or to better store them (Hemmelskamp, 1997, p. 178). Because end-of-pipe technologies focus on treating after-the-fact (effluent) waste streams, they do not reflect a prudent environmental strategy. In contrast, production-integrated technologies save natural resources via pollution prevention or resource recovery and thus support a sustainable development. The concept "sustainable development" has different aspects. Among others, they include causal-oriented protection of the environment which is ensured by production-integrated measures. Thus, in order to pursue the goal of a sustainable development a hierarchy of technological measures can be established. Process technologies that reduce waste (water) at source should have first priority. The second priority is resource recovery and the last priority is end-of-pipe treatment of after-the-fact (effluent) waste streams (Fuller, 1999, p.6 and pp. 86-87).

3.2 Patent classes and keywords

In order to get data about patent activities of chemical firms we used two different methodologies.

End-of-pipe technologies are recorded by the German Patent Office in the class C02F of the International Patent Classification (IPC): Treatment of water, waste water, sewage or sludge. This class is further divided into six subclasses:

C02F - 1/00: treatment of water, waste water, or sewage (3/00 to 9/00 take precedence)

C02F - 3/00: biological treatment of water, waste water, or sewage

C02F - 5/00: softening of water, preventing scale; adding scale preventatives or scale removers to water, e.g. adding sequestering agents

C02F - 7/00: aeration of stretches of water

C02F - 9/00: multistage treatment of water, waste water or sewage

C02F - 11/00: treatment of sludge; devices therefore.

As already mentioned, production-integrated technologies can be further subdivided into technologies that reduce waste water at source and recycling technologies (see figure 1). For both subclasses of technologies there is no specific patent class because processes are classified according to their technical characteristics, rather than their environmental effects. Data about *recycling* technologies that reduce water pollution are obtained by using the following keywords in patent analysis: recycl* (recycle), wiedergew* (reclaim), wiederverw* (re-use), rezyklier* (recycle), zurückf* (return), rückf* (return) and rückg* (recover). We used a variety of keywords to cover a broad range of recycling technologies and to account for the fact that different firms may use different terms within the main claims of the patent application.[1] In order to avoid double or triple counting of patent applications we checked the number of every application. The keywords did not allow to identify whether the selected recycling technologies *directly* reduce water pollution or not. However, recycling technologies that directly reduce emissions into air and soil may *indirectly* reduce water pollution. For example, less emissions into air will result in less washing water because end-of-pipe treatment of these emissions will be avoided. Or, at a national or international level, less emissions into air reduce the pollution of water resources through rainfalls containing harmful substances. Thus, recycling technologies may indirectly reduce water pollution.[2] In this sense, the number of recycling technologies calculated by us is an *upper limit* for recycling technologies that reduce water pollution.

The distinction between end-of-pipe and production-integrated technologies should not be considered as totally dichotomous (Kemp, 1997, p.12). In our study it also turned out that some process technologies which we identified as recycling technologies were classified as waste water treating end-of-pipe technologies (C02F). In order to avoid double counting we subsumed these cases under the class of end-of-pipe technologies.

Data about patent applications in the field of production-integrated technologies that reduce waste water at source could not be obtained because the main claim of the patent application does not indicate that emissions are avoided at source. Hence, the use of keywords was not possible. Figure 2 takes up figure 1 and depicts the classes of patents and keywords we used to obtain data on patent activities at the firm level.

Figure 2. Classes of patents and keywords to measure R&D that reduces water pollution.

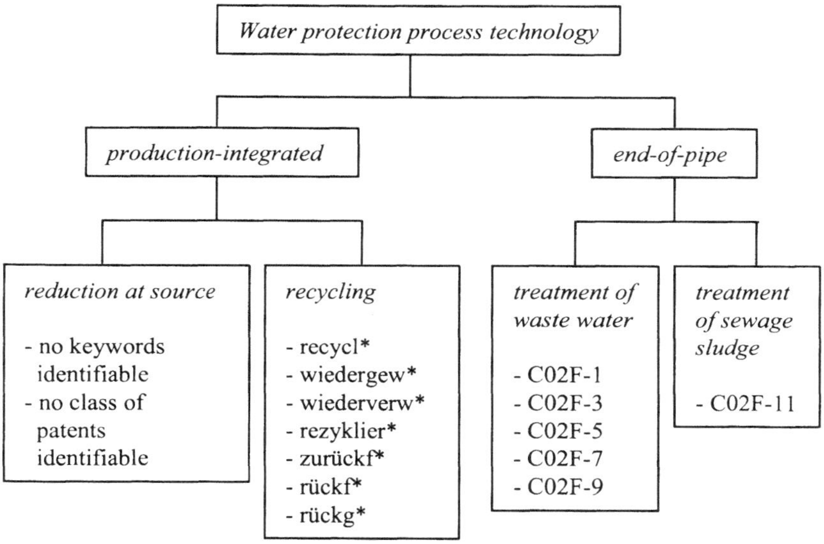

Source: Deutsches Patentamt, 1995, pp.20-21; own keywords

4. DESCRIPTION OF THE SAMPLE OF FIRMS

To carry out a patent analysis for the whole population of German chemical firms would be too time consuming. Therefore, we decided to take a sample including a cross-section of the chemical industry of West Germany (FRG) but exclude chemical firms of East Germany (former GDR). For firms of both regions were operating with capital stocks that differ substantially with respect to age and composition. For instance, the production of chemical substances was based on brown coal in the former GDR, whereas it was based on oil in West Germany (Wirtschaftspolitische Diskurse Nr. 19, 1991). Furthermore, West German firms had to account for many environmental regulations when developing and operating new plants. Therefore, the knowledge base of firms of both regions was quite different before 1990. The existing knowledge base of firms in turn influences their capability to discover new technologies. Hence, for firms of both regions, the competitiveness and the capability to discover new processes was fundamentally different *even after* German Unification. Eight years after unification the structures of the two chemical sectors still differed significantly (see table 1).

Table 1. Economic performance of West and East German chemical firms in 1998

German region	*Employees*	*Employees per firm*	*Turnover (Euro) per firm*	*Share of turnover by foreign trade*
West Germany	465,177	412	243,029	47.1
East Germany	26,923	177	169,521	27.7

Source: Statistisches Bundesamt, 1999, pp.177, 261, 273, 309, 405

The selection of chemical firms is based on the company data base of the "Verband der Vereine Creditreform (VVC)". VVC is the largest credit ranking agency in Germany. According to the data base of VVC, the chemical sector in West Germany contained 1698 chemical firms having more than 19 employees in 1998 (see table II, second column). In contrast, the German Federal Statistical Office (Statistisches Bundesamt) counts only 1127 West German chemical firms with more than 19 employees (Statistisches Bundesamt, 1999, p.261). The rather large difference may be due to two factors: Firstly, about 10 percent of the chemical firms which we selected by means of the VVC data base do not regard themselves as belonging to the chemical industry (information was obtained by telephone interviews). Secondly, the German Federal Statistical Office classifies firms according to their *main* business activities in order to avoid double counting (Fachserie 4, Reihe 4.1.1, 1999, p.10). In contrast, by using the VVC data base we account for firms with at least one business activity which belongs to the chemical sector irrespective of its relative importance.

In order to select the individual firms for the empirical investigation we classified the population of chemical firms into four groups according to the number of employees (see table 2, first column).

Table 2. Classified sample of West German chemical firms in 1998

Firm size	*Total of West German chemical firms*	*Selected firms*	*Adjusted sample*
5000 and above	20	20	18
500-4999	174	20	19
50-499	844	20	16
20-49	660	20	19
Total	1698	80	72

Source: Data base of the VVW, randomised selection

Within each class we chose 20 firms randomly (within the class with more than 4999 employees the firms were selected with a probability of one). Hence the sample contains 80 chemical firms of different firm sizes. The idea behind that method of selection was twofold: Firstly, establishing different classes of firm size is a precondition for analysing the invention-firm size relationship Secondly, by choosing at least the 20 largest chemical firms we cover a fairly large part of turnover and thus production and water pollution of the chemical industry because the business concentration within the chemical sector is very high. Thus, the 25 largest firms (with respect to turnover) count for more than 50 percent of turnover of the overall sector (Statistisches Bundesamt, 1998, p.28). Hence, the largest share of chemical substances will be produced by a small number of large firms. This is due to *economies of scale* in production and economies of scope in innovation in industrial chemistry. Economies of scale in production arise because the equipment of chemical plants primarily consists of pipes and cylinders. Thus capital costs depend on surface area of pipes while capacity depends on volume of pipes (Freeman and Soete, 1997, p.98). Hence - as a rule of thumb - if capacity increases capital costs increase only by about one half.

By means of telephone interviews we adjusted our sample (see table 2, fourth column). It turned out that some firms belonged to another class of firm size than originally classified. Furthermore, 8 out of 80 firms did not consider themselves as chemical firms.

5. EMPIRICAL RESULTS

5.1 Inventive activities in the field of end-of-pipe technologies

Among the 72 chemical firms of our sample the very large firms, including among others the chemical giants BASF, Bayer and Aventis (former Hoechst), count for nearly all patent applications in the 1990s (see table 3, last column).

Table 3. Patent applications of chemical firms in different fields of end of pipe technologies that treat waste water and sewage sludge (1991-1998)

Firm	*C02F -1*	*C02F -3*	*C02F -5*	*C02F -7*	*C02F -9*	*C02F -11*	*Total*
Henkel, Düsseldorf	17	3	4	-	1	1	26
BASF, Ludwigsh.	20	-	1	-	-	2	23
Bayer, Leverkusen	14	1	2	-	2	3	22
Hoechst, Frankfurt* including Aventis**	9	5	-	-	1	1	16
Solvay, Hannover***	11	1	-	-	4		16
Degussa, Frankfurt	12	-	-	-	-	-	12
Hüls, Marl	7	-	2	-	-	-	9
Agfa-Geyaert, Leverkusen	2	-	-	-	-	-	2
Schering, Berlin	1	-	-	-	1	-	2
Wacker, Burghausen	1	-	-	-	-	-	1
Ciba Specialty, Basel	-	-	-	-	-	1	1
500-4999							
Kronos-Titan, Leverkusen	1	-	-	-	-	-	1

* former Hoechst, Frankfurt (including among others Aventis Pharma, Frankfurt, InfraServ, Frankfurt and parts of actual Clariant, plus the central R&D department) in the years 1991-1997

** Aventis, Frankfurt (including among others actual Aventis Pharma, Frankfurt, Infraserv, Frankfurt and Aventis Research & Technologies (former central R&D department of Hoechst) in 1998

*** including Solvay Umweltchemie, Hannover; Solvay Alkali, Hannover; Solvay Flour und Derivate, Hannover

Source: Own patent analysis

Hence, our data suggest that in particular the largest chemical firms with 5,000 and more employees develop end-of-pipe technologies and apply for patents in the field of end-of-pipe technologies. Since the 1960s firms like BASF, Bayer and Aventis have implemented many waste water treatment plants in various establishments and have accumulated knowledge about technological processes that reduce waste water (Bayer-Umweltperspektive, 1987, p. 22). In particular in the late 1970s and 1980s these firms developed radical end-of-pipe innovations to reduce industrial waste water. An important example is the "tower biology" (Turmbiologie) developed by Bayer (Bayer-Umweltperspektive, 1987, pp. 21-26). This knowledge base positively affected waste water reducing discoveries in the 1990s.

Most patent applications are classified within the field C02F-1 which contains all those patent applications that cannot be assigned to the classes C02F-3 to C02F-11. And within C02F-1, the patent applications of the chemical firms especially referred to the treatment of water, waste water and sewage by (1) oxidation, (2) removing specified dissolved compounds (e.g. heavy metal compounds), (3) extraction, (4) sorption, (5) flocculation or precipitation of suspended impurities, and (6) iron-exchange.

The very large chemical firms were responsible for most of the patent applications in the field of end-of-pipe technologies. However, except Solvay, these patent applications were only a minor fraction of all patents that were applied by these three firms between 1991 and 1998 (see table 4, column 5).

Table 4. The importance of patent applications in the field of water benign process technologies for the largest chemical applicants (1991–1998)

Firm	*Number of patent applications: end-of-pipe technologies (1)*	*Number of patent applications: recycling technologies (2)*	*Total number of patent applications (3)*	*(1) / (3) in (%)*	*(2) / (3) in (%)*
Henkel	26	21	2769	0.94	0.76
BASF	23	105	5195	0.44	2.02
Bayer	22	69	4378	0.50	1.58
Hoechst* Aventis**	16	40	2612	0.61	1.53
Solvay***	16	7	230	6.96	3.04

* former Hoechst, Frankfurt (including among others actual Aventis Pharma, Frankfurt, InfraServ, Frankfurt and parts of actual Clariant, plus the central R&D department) in the years 1991-1997

** Aventis, Frankfurt (including among others actual Aventis Pharma, Frankfurt, Infraserv, Frankfurt and Aventis Research & Technologies (former central R&D department of Hoechst) in 1998

*** including Solvay Umweltchemie, Hannover; Solvay Alkali, Hannover; Solvay Flour und Derivate, Hannover

Source: Own patent analysis

5.2 Inventive activities in the field of recycling technologies

As already mentioned, data on patent applications in the field of recycling are obtained by using keywords. The numbers calculated by us can be regarded as upper limits *for recycling technologies that reduce water pollution* (see section 3). As already mentioned, data on process technologies that reduce waste water at source could not be obtained because the main claim of the patent application does not indicate whether emissions are avoided at source or not. Thus, the use of keywords was not possible.[3]

Again, within our sample, the largest firms count for nearly all patent applications in the field of recycling technologies in the 1990s (see table 5).

Table 5. Patent applications of chemical firms in the field of recycling technologies (1991-1998)

Firm	*Number of patent applications: recycling technologies*
BASF, Ludwigshafen	105
Bayer, Leverkusen	69
Hoechst*, including Aventis**, Frankfurt	40
Henkel, Düsseldorf	21
Degussa, Frankfurt	17
Hüls, Marl	17
Solvay, Hannover***	7
Merck, Darmstadt	3
Schering, Berlin	3
Wacker, Burghausen	2
500-4999	
Lohmann, Neuwied	2
B.Braun Melsungen, Melsungen	1
Kronos-Titan, Leverkusen	1

* Former Hoechst, Frankfurt (including among others actual Aventis Pharma, Frankfurt, InfraServ, Frankfurt and parts of actual Clariant, plus the central R&D department) in the years 1991-1997

** Aventis, Frankfurt (including among others actual Aventis Pharma, Frankfurt, Infraserv, Frankfurt and Aventis Research & Technologies (former central R&D department of Hoechst) in 1998

*** including Solvay Umweltchemie, Hannover; Solvay Alkali, Hannover; Solvay Flour und Derivate, Hannover

Source: Own patent analysis

However, the inventive activities of the three giants BASF, Bayer and Aventis (former Hoechst) are outstanding within the class of very large firms. These companies have central R&D departments employing several hundreds of natural scientists, engineers and laboratory personnel to generate new recycling processes.

It is striking that the inventive activities of very large pharmaceutical firms such as Merck and Schering are very low in the field of recycling technologies. The core business of these pharmaceutical firms is product rather than process innovation. Therefore, it is likely that they refrain from process inventions.

Our data indicate that the proportion of water benign inventions in the total of corporate inventive activities is higher in the field of end-of-pipe technologies than in the field of recycling technologies for Henkel, and in particular for Solvay (see table 4, column 5 and 6).

In contrast, for the three largest patent applicants BASF, Bayer, and Hoechst (and also for Degussa and Hüls) patent applications in the field of recycling technologies play a larger role within the R&D portfolio than patent applications in the field of end-of-pipe technologies (see table 4, column 5 and 6). One important reason is that the optimisation of production processes (including the recovery and exploitation of joint products) to reduce production costs has played an important role since the beginning of industrial chemistry. Chemical firms accumulated a lot of knowledge in the optimisation of production processes. Relatively new to chemical firms is just the *direction* of optimisation towards the protection of environmental resources such as water. BASF is an outstanding example for the optimisation of production processes (see box 1).

Box 1: BASF and Its "Verbund"

The major establishment of BASF at Ludwigshafen contains about 350 individual factories that produce about 8000 different chemical and pharmaceutical products. The individual factories are linked together with about two thousand kilometres of pipes and make intensive use of their joint products. Basic chemicals or intermediate products of different factories are processed by other factories on the firm area. Furthermore, the heat jointly produced during production processes provides the energy for other production processes. The described integration of manufacturing intermediate and final chemicals is an important part of the philosophy of BASF that dates back to its founding by Friedrich Engelhorn in 1865. The philosophy is called "Verbund". The extent of this integration at BASF is unique in the chemical industry and the term "Verbund" has virtually developed as an international brand name for BASF. Via its "Verbund"-system at Ludwigshafen BASF saves production costs of about 500 millions Euro (p.a.) and reduces carbon dioxide emissions by about two and a half million of tons per year. As the "Verbund"-system furthers BASF's competitiveness, its continuous optimisation is an important objective of the BASF.

Source: Andreas Kempf, 2000, p.13; BASF Gruppe 1999

6. CONCLUSIONS

This chapter sought to examine how firm size relates to inventions in water benign process technologies during the 1990s in West Germany.

In order to analyse this question we randomly selected 80 firms out of a classified population of chemical firms. After adjustment for coverage, our sample contained 72 firms.

We measured corporate inventive activities with by means of patent applications. In order to get data about patent applications in the field of water benign process technologies we selected the patent class C02F and used keywords.

Our empirical results indicate that with respect to both end-of-pipe and recycling technologies especially very large chemical firms predominate invention. With respect to recycling technologies the outstanding inventive activities of the three giants BASF, Bayer, and Aventis are striking.

Except Solvay and Henkel, within the portfolio of patent applications of large firms, patent applications in the field of end-of-pipe technologies play a less important role than patent applications in the field of recycling technologies. Several reasons for that invention behaviour can be given. First, the optimisation of production processes (including the recovery and exploitation of joint products) in order to reduce production costs has played an important role since the beginning of industrial chemistry. Chemical firms accumulated a lot of knowledge in the optimisation of production processes. Second, a part of the firms favours recycling technologies at the expense of end-of-pipe technologies because of their environmental awareness. Third, firms favour to invent and implement recycling technologies at the expense of end-of-pipe technologies as governmental emission standards get tighter and tighter (Becker and Englmann 2003). The ongoing addition of end-of-pipe technologies due to tightened standards during the 1990s in Germany requires ongoing material input and energy consumption while the incumbent production process remains unchanged. Thus, the implementation of (newly invented) end-of-pipe technologies tend to reduce productivity and thereby competitiveness.

NOTES

[1] In order to select valid keywords Grupp proposes to include linguistics in research in order to enlarge knowledge about how keywords are developed and accepted in scientific circles (Grupp, 1999, p.27).

[2] The authors are aware of the fact that some recycling technologies may reduce air pollution at the expense of water pollution.

[3] Data on invention of integrated technologies that reduce water pollution at source was obtained via a postal survey which we carried out in Spring 1999. Altogether, 31 out of the 72 firms above (see table II)

responded to our questionnaires. The firms were asked to refer their replies on one establishment. In particular establishments with at least 500 employees implemented integrated technologies between 1996 and 1998 which were (at least partially) the result of in-house R&D activities.

7. REFERENCES

Anastas, P. T., & Williamson T. C. (1998a). Frontiers in Green Chemistry: In P. T. Anastas & T. C. Williamson (Eds.), Green Chemistry: Frontiers in Benign Chemical Syntheses and Processes (pp. 1-26). Oxford: Oxford University Press.

Anastas, P. T., & Warner J. C. (1998b). Green Chemistry: Theory and Practice. Oxford: Oxford University Press.

Bayer (1987). Die Bayer-Umweltperspektive. Leverkusen: Bayer.

BASF Gruppe (1999). Der Verbund. In BASF Gruppe. Retrieved May 16 2003, from http://www.basf.de/de/corporate/verbund/.

Becker, F., & Englmann, F. C. (2003). Public Policy, Voluntary Initiatives and Water Benign Process Innovations: Empirical Evidence from the West German Chemical Industry during the Mid-1990s: In K. M. Weber & J.Hemmelskamp (Eds.), Towards Environmental Innovation Systems (pp. ??-??). Heidelberg: Springer.

Dosi, G. (1982). Technological Paradigms and Technological Trajectories. Research Policy, 11(3), 147-162.

Freeman, C., & Soete, L. (1997). The Economics of Industrial Innovation (3rd ed.). London: Pinter.

Grupp, H. (1998). Foundations of the Economics of Innovation. Cheltenham: Edward Elgar.

Grupp, H. (1999, April), Evaluation von Fördermaßnahmen mit Leistungsindikatoren. Paper presented at the meeting Innovationsförderung: Möglichkeiten der empirischen Effizienzkontrolle, Leipzig.

Hemmelskamp, J. (1996). Umweltpolitik und Innovation. ZEW-Discussion Paper, 96-23. Mannheim.

Hemmelskamp, J. (1997). Environmental Policy Instruments and their Effects on Innovation. European Planning Studies, 5(2), 177-193.

Kempf, A. (2000, October 28). Selbst die US-Konkurrenz spricht von "Verbund". Stuttgarter Zeitung, p. 13.

Müller-Fürstenberger, G. (1995). Kuppelproduktion. Heidelberg: Physica Verlag.

Pavitt, K., Robson, M., & Townsend, J. (1987). The Size Distribution of Innovating Firms in the UK: 1945-1983. The Journal of Industrial Economics, 35(3), 297-316.

Särndal, C. E., Swensson B., & Wretman J. (1992). Model Assisted Survey Sampling. New York: Springer.

Scherer, F. M. (1965). Firm Size, Market Structure, Opportunity, and the Output of Patented Inventions, American Economic Review, 55, 1097-1123.

Statistisches Bundesamt (1998). Fachserie 4, Reihe 4.2.3 Konzentrationsstatistische Daten für das Verarbeitende Gewerbe, den Bergbau und die Gewinnung von Steinen und Erden sowie für das Baugewerbe 1995 und 1996. Wiesbaden.

Statistisches Bundesamt (1999). Fachserie 4, Reihe 4.1.1 Beschäftigung, Umsatz und Energieversorgung der Unternehmen und Betriebe im Bergbau und im Verarbeitenden Gewerbe 1998. Wiesbaden.

Schumpeter, J. A. (1993). Kapitalismus, Sozialismus und Demokratie (7th ed.). Tübingen: Francke.

Friedrich Ebert Stiftung (1991). Chemiestandort Ostdeutschland. Wirtschaftspolitische Diskurse 19. Bonn.

SUBJECT INDEX

GPSR Compliance

The European Union's (EU) General Product Safety Regulation (GPSR) is a set of rules that requires consumer products to be safe and our obligations to ensure this.

If you have any concerns about our products, you can contact us on ProductSafety@springernature.com

In case Publisher is established outside the EU, the EU authorized representative is:

Springer Nature Customer Service Center GmbH
Europaplatz 3
69115 Heidelberg, Germany

Batch number: 10816040

Printed by Printforce, the Netherlands